Get on the Air with HF Digital

The Beginner's Guide to PSK31, RTTY and More!

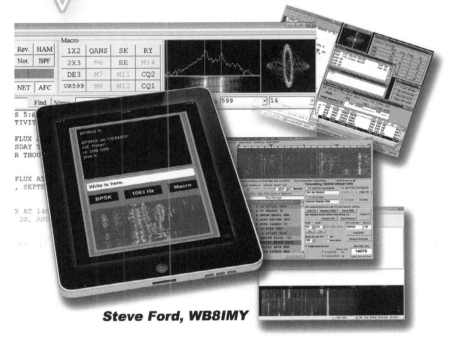

Steve Ford, WB8IMY

Production Staff: **Maty Weinberg**, KB1EIB, Editorial Assistant

David Pingree, N1NAS, Senior Technical Illustrator:

Carol Michaud, KB1QAW, Technical Illustrator

Jodi Morin, KA1JPA, Assistant Production Supervisor: Layout

Sue Fagan, KB1OKW, Graphic Design Supervisor: Cover Design

Michelle Bloom, WB1ENT, Production Supervisor: Layout

ARRL The national association for **AMATEUR RADIO**®

225 Main Street, Newington, CT 06111-1494

www.arrl.org

Contents

Foreword

Interest in amateur HF digital communications is growing at a rapid rate. On any given day, even when propagation conditions are poor and the bands are practically "closed," you can hear the sounds of digital conversations taking place.

The fact that most amateurs own computers is part of what is driving the popularity of HF digital. The other major factor is that so many hams live in homes that suffer severe restrictions on the kinds of outdoor antennas that can be installed – if any. Digital modes allow these amateurs to communicate with low power and compromiseD antennas, letting them enjoy Amateur Radio when they'd otherwise be off the air entirely.

The only problem is that HF digital technology is a foreign concept to a substantial number of operators. They want to try it, but don't know where to start. That's the purpose of *Get On the Air With HF Digital*.

This book offers clear, often step-by-step instructions about how to set up an HF digital station – everything from attaching cables to configuring software. Even though software changes over time, the instructions provided in this book are likely to apply across several software versions.

Get On the Air With HF Digital emphasizes the hands-on approach. The goal is to give you all the advice you need to take to the airwaves as quickly and easily as possible.

David Sumner, K1ZZ
ARRL Chief Executive Officer
2011

Chapter One

Let's Build an HF Digital Station

The Short Scoop
An HF digital station boils down to three essential pieces: a
radio, a computer and a device that ties them together.

The HF Transceiver

The radio requirements for HF digital are surprisingly straightforward. You don't need to run out and purchase a special radio with sophisticated features. All you need is an SSB voice transceiver. *Period.*

Any transceiver made within the last 20 years will work well with most HF digital modes. Frankly, even very old SSB transceivers can be pressed into service.

When considering an older radio, however, there are two things to keep in mind:

Stability—Older transceivers, especially radios that use vacuum tubes, may tend to drift in frequency. Drift is deadly to digital communications. If you must use an old rig as your HF digital transceiver, you may need to allow it to warm up for as long as 30 minutes prior to operating.

Transmit/Receive Switching Speed—One digital mode known as PACTOR requires the radio to jump from receive to transmit (and back) in milliseconds. Many older radios can achieve this, but it is hard on their switching relays. I once managed to use an ancient Drake TR-4 transceiver to make a PACTOR contact, but it sounded as though the radio was about to shake itself to pieces! Newer radios with solid-state switching are better for this application.

If your radio dates from about 1990 to the present day, you'll likely be in fine shape. These rigs are stable and most include all the features you'll need for HF digital. The singular exception involves some of the inexpensive SSB transceivers designed for low-power (QRP) operating. These rigs aren't usually intended for

I once managed to make a PACTOR contact with an old Drake TR-4 transceiver, but the ancient radio's relays were working overtime!

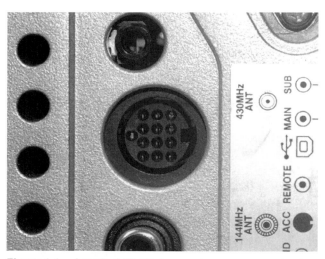

Figure 1.1 – A typical 13-pin transceiver accessory jack. Among the pins of most interest to HF digital operators are those that carry receive audio, transmit audio and the PTT (Push To Talk) line.

digital applications, so their stability may be questionable. They may also not offer connections to external devices through jacks known as *ports*.

Radio Ports

Modern radios get along surprisingly well with computers and other external devices. In fact, most modern HF transceivers are designed with external devices in mind. They offer a variety of ports depending on the model in question.

Nearly every HF rig manufactured within the past decade includes an "accessory" port of some kind. Typically these are multipin jacks (as many as 13 pins) that provide connections for audio into and out of the radio, as well as a pin that causes the radio to switch from receive to transmit whenever the pin is grounded. This is often called the *PTT* or *Push To Talk* line. Some manufacturers also call it the "Send" line. Take a look at the typical accessory jack shown in **Figure 1.1**.

These accessory ports are ideal for connecting the kinds of interface devices we use to operate HF digital. In addition to the PTT function, accessory ports provide receive audio output at *fixed* levels that never change no matter where the VOLUME knob is set. This is a highly convenient feature that you'll appreciate when operating late at night after everyone has gone to bed. You can turn the VOLUME knob to zero and still have all the receive audio you need!

Be aware that some radio manufacturers label accessory ports as "data" or "digital" ports. This causes no end of confusion because modern rigs often offer two types of connections: a true accessory port with audio and transmit/receive keying lines and another port that allows a computer to actually control the radio. The confusion occurs when hams attempt to figure out which ports they need to use. Let's clear the air!

For the purpose of getting on the air with HF digital, the only type of radio "control" we care about is the ability to switch from receive to transmit and back

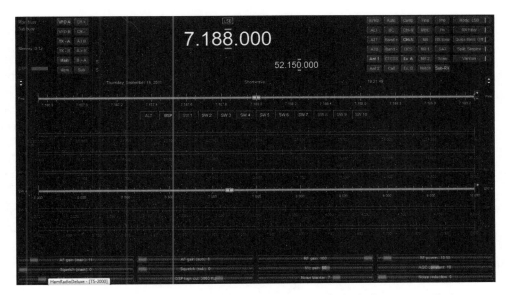

The CAT screen from *Ham Radio Deluxe*. This software is typical of programs that can control almost all aspects of a transceiver, but this kind of control isn't necessary for HF digital.

again. That connection is available at the *accessory port*, even though the port may go by a different name.

The kind of control the transceiver manufacturers have in mind goes way beyond the act of simply switching between transmit and receive. They are talking about the computer taking over almost every function of the radio; that's a different animal entirely. Full computer control usually involves software that does many things such as displaying and changing the transceiver's frequency, raising and lowering the audio level, scanning memory frequencies and a great deal more. Some computer control software is so elaborate that the radio itself can be placed out of sight and all control conducted at the keyboard and monitor screen. Many amateurs use this capability to control their rigs remotely over the Internet at great distances.

Transceivers have separate ports for this type of computer interfacing, often referred to as *CAT*, or Computer Aided Transceiver, and these are most definitely are *not* accessory ports. On the contrary, they are ports strictly designed to swap data with external computers. They come in several varieties...

TTL: Transistor-Transistor Logic. These ports require a special interface to translate the serial communication from your computer to TTL pulses that your radio can comprehend.

USB: Universal Serial Bus. Although the consumer electronics world adopted USB years ago, transceiver manufacturers have been somewhat slower to catch on.

RS232: This is a serial port that can be connected directly to your computer if your computer has a serial (COM) port. Most modern computers have done away with serial ports, but you can use a USB-to-serial converter to bridge the gap.

Ethernet: This port allows the transceiver to become a "network device," just like your wireless router, printer, etc. Only a handful of transceivers offer Ethernet ports at this time.

If you want to set up complete computer control of your radio, we'll discuss some options that will allow you to do this. But to be perfectly clear and blunt, let me say this: *To get on the air and enjoy HF digital, you do* not *need full computer control and you do not need to concern yourself with this transceiver port. The accessory jack (or whatever your radio manufacturer calls it) is the only port that matters.*

What? No Radio Ports?

What if your SSB transceiver doesn't have an accessory port? No problem. You can use the microphone jack as your connection for transmit/receive switching as well as the audio input. For the audio output, you can use the external speaker or headphone jack. This isn't an ideal situation, but it works. In fact, many HF digital operators take this approach.

Typical microphone and headphone jacks.

Duty Cycle

When talking about your transceiver, we need to spend a little time discussing the concept of *duty cycle*. A somewhat crude definition of duty cycle is the time that a radio spends generating RF output as a fraction of the total time under consideration. In HF digital terms, think of duty cycle as the amount of time your radio is generating RF during any given transmission compared to the amount of the time during the same transmission when RF output falls to zero. A 100% duty cycle would mean that your radio is cranking out RF continuously throughout the entire transmission; the RF output level never falls to zero.

Now you may say, "I guess I'm always using a 100% duty cycle. After all, my radio is always generating RF whenever I'm transmitting."

Not necessarily.

When you are transmitting digital, CW or even SSB voice, your rig may not be operating at a 100% duty cycle. Consider SSB voice as an example. Whenever you speak into the microphone, the RF output level changes dramatically as your voice changes. It can go from 100% output to zero in a fraction of a second. The same is true for CW. Whenever your CW key is open between the dots and dashes, your transceiver output is at zero.

Measured over a period of time (your transmitting time), the duty cycle of SSB voice is actually about 40%; CW is often as low as 30% or even less if you are a particularly slow sender.

HF digital modes also vary in duty cycle. Some modes such as radioteletype (RTTY) push your radio to a duty cycle of nearly 100%. Others result in much lower duty cycles.

So why should you care about your duty cycle?

The answer is that your radio may not be designed for the type of punishment a high duty cycle transmission can inflict. When you operate your radio at a 100% duty cycle, you are demanding that its final amplifier circuits produce the full measure of output—whatever you've set that output level to be – for the entire time you are transmitting. The result is heat, and potentially a lot of it. Apply enough heat to a circuit for a sufficient length of time and you'll see components begin to fail, sometimes in spectacular fashion.

Some manufacturers don't consider the possibility that their SSB voice transceivers might be pressed into service as digital transceivers. They design the radios to tolerate the duty cycle of a typical voice transmission. If you use this same radio to enjoy a high duty cycle digital mode such as RTTY, you could be asking the radio to operate well outside its design limits—with unfortunate results.

Always read your transceiver manual before attempting to use the radio for digital operating. The manufacturer may advise you to reduce the RF output by as much as 50% when using high duty cycle modes. This keeps the heat generation manageable. When in doubt, or when you notice that your radio is becoming particularly hot, reduce the RF output. You'll find that digital modes don't require a great deal of output power anyway, so chances are you won't notice the 50% reduction.

Transceiver Filters

Another important transceiver feature to consider is receive filtering. In most cases we're talking about the filters located in the Intermediate Frequency (IF) stage of your radio. These can be physical filters; modules that plug in or are soldered into the radio's circuit board. Many modern transceivers use Digital Signal Processing (DSP) at the IF stage rather than physical filters. The advantage of DSP is that it is often designed to be continuously variable. This means that you can narrow or expand the filter with the push of a button or the twist of a knob.

As you'll learn later, most popular HF digital modes use wide receive-audio bandwidths for reception, so it isn't strictly necessary to have an IF filter narrower than the typical SSB voice bandwidth of about 2.8 kHz. All SSB transceivers meet this requirement.

There are some important exceptions, though. If you decide to try your hand at RTTY contest operating, you'll quickly discover that a 2.8 kHz IF bandwidth is entirely too wide. You won't be able to easily separate individual signals in a sea of RTTY contest chaos. In this environment you need to narrow the IF bandwidth to at least 500 Hz. In ultra-crowded conditions you may need a bandwidth as narrow as 300 Hz.

There may be other situations where you'll want to use a narrower IF bandwidth. Let's say you are trying to communicate with a weak station and a much stronger station begins transmitting a rock-crushing signal about 1 kHz up the band. Your radio's Automatic Gain Control (AGC) is going to respond by dropping your receiver sensitivity into the basement. The weak station will become weaker still, or disappear altogether.

It isn't practical to ask the gigawatt station to move, so your only alternative is to dramatically narrow your IF bandwidth to the point where his signal is eliminated and only the weak signal remains. Continuously adjustable IF DSP filtering is terrific for this application. You can create an IF filter as narrow as necessary and put it right on the signal you are trying to hear. With ultra-sharp DSP filters, everything outside the passband is effectively gone.

If your transceiver already has selectable or variable IF filters, you're all set. If not, consider adding a 500 Hz IF filter if your radio will allow you to do so. It could cost you as much as $130, but it is a good investment in the future, especially for contesting.

AFSK vs FSK

Here is an interesting point that often causes consternation among digital operators when they consider their HF transceivers.

Basic HF SSB transceivers typically offer at least three operating modes: Upper Sideband (USB), Lower Sideband (LSB) and CW. Others include AM and even FM. For HF digital operation USB or LSB is all you need.

That said, you'll also notice that some HF rigs include a digital mode selection that may be labeled "RTTY," "Digital" or "Data." Here is where the gnashing of teeth begins because the way the radio behaves when you select this mode can vary depending on the design. In other words, these labels can mean different things on different radios.

For instance, the "Digital" mode may simply control how the IF filters can be used. The radio may forbid you from selecting a narrow IF filter when you are in one of the SSB modes, but it will suddenly relent and allow you to access a narrow filter when you enter the "Data" mode. It may also include a function that "looks" for the incoming audio at the accessory jack whenever you select the "Data" mode. Otherwise it ignores the incoming audio at the accessory jack entirely. This is odd design behavior, but it happens.

What is most important for our discussion, however, is the fact that some manufacturers use this mode label to indicate that the radio will switch to *FSK*. What does this mean?

The vast majority of HF digital

The CW/RTTY mode button on an ICOM transceiver. But is it AFSK RTTY (and other digital modes), or FSK RTTY alone? Time to consult the user manual!

operators use Audio Frequency Shift Keying or *AFSK*, although they may not realize it. The audio tones from their sound devices are applied to their radios and converted to shifting RF frequencies at the output, hence the term *Audio Frequency Shift Keying*.

But there is yet another way to create shifting RF. You can take data pulses (not sound) directly from the computer and use those pulses to directly shift a transceiver's master oscillator from one RF frequency to another. Since there is no audio involved, this is known as Frequency Shift Keying or *FSK*.

Is there an advantage to using FSK vs AFSK? Not really. It is mostly a matter of convenience. Some HF digital transceivers will not allow you to use narrow IF filters unless you operate using the FSK mode (this design quirk is in decline, though).

The problem with FSK in amateur transceivers is that it is limited to shifting between only two frequencies. Many amateur digital modes use more than two frequencies. PSK31 – the most popular digital mode—doesn't shift frequency at all; it shifts *phase* instead. The only common digital mode that can benefit from true FSK is RTTY since RTTY signals only shift between two frequencies.

FSK as it exists in Amateur Radio transceivers is a feature for RTTY aficionados alone. The irony is that it is impossible to tell the difference between a properly modulated AFSK RTTY signal and a RTTY signal generated by FSK.

So why bother? Again, the answer is convenience—the FSK RTTY operator never has to worry about sending too much audio to his radio and at the same time he is assured of having full access to all the IF filters that might otherwise be missing if he weren't using this mode.

If your transceiver offers a "RTTY," "Data" or "Digital" mode, read your manual carefully and find out exactly what this means. Chances are that "RTTY" means FSK. "Data" or "Digital" likely means AFSK, although even the manufacturers confuse the terminology. You'll find some manuals referring to AFSK as FSK. The authors are not wrong, strictly speaking. After all, regardless of whether you're using AFSK or FSK the RF signal frequencies at the output are still shifting back and forth. It is "frequency shift keying" in either case. The difference is in the method you use to create the shifty signals.

If you're unsure, look at the hookup diagram in the manual. Does it show digital transmit audio being applied to the radio at the accessory or microphone jack? If so, it is AFSK. When you select the "Digital" or "Data" mode, you're really in the SSB mode, but using audio to generate the frequency changes.

But if the manual shows keying data being supplied to an "FSK" line, it is indeed true FSK.

Computers

As this book was being written, the computer world was entering a state of flux. By "flux" I mean rapid change.

From the early 1980s through about 2005, the desktop computer was king among ordinary consumers and Amateur Radio operators. This is a computer in a separate, stand-alone case connected to a monitor screen, keyboard and mouse. Inside the computer case there is a sound device of some sort, either a dedicated *sound card* or a set of sound-processing chips on the motherboard. The computer connects to peripheral devices through the use of serial (COM) ports.

From 2005 onward we saw two important changes. One was the fact that laptop computers became more powerful and affordable. The other was that serial ports disappeared in favor of USB ports in both laptops *and* desktop machines.

By 2010 laptops (and to a lesser degree, netbooks) began to dominate Amateur Radio stations. As hams upgraded their computers, they no longer saw the need for bulky desktop systems when sleek laptops would do quite nicely.

As this book went to press, even laptops were facing stiff competition from *tablet* computers such as the Apple iPad. In Amateur Radio stations, laptops and desktops are still the most popular computers, but this is likely to change as more ham applications are developed for tablets.

At the time this book was written, desktop PCs were fading in the consumer world. Among hams, however, they were still widely used.

For now, however, our focus will remain primarily on laptops and desktops. With that in mind, what kind of laptop or desktop do you need for HF digital operating?

The good news is that most HF digital software does not require powerful computers. Any ordinary off-the-shelf consumer-grade computer will do the job. If you are buying new, don't overspend for a powerful computer you won't need. Frankly, a decent used computer that's only a couple of years old will be more than adequate.

If you're thinking about a small netbook, be careful to check the specifications. Some netbooks don't include audio ports and as you'll see shortly, audio ports are highly important.

In terms of operating systems, most HF digital software is written for Microsoft *Windows*. Many of these programs were created during the *Windows XP* era, but they run well on both *Windows Vista* and *Windows 7*. Some amateurs have reported problems with older software running under 64-bit *Windows 7*, but in my experience this problem is uncommon.

There is HF digital software available for *MacOS* as well, although not as much variety. *Linux* users will find a number of HF digital applications, too.

In many ham stations, laptops are taking over from desktop computers.

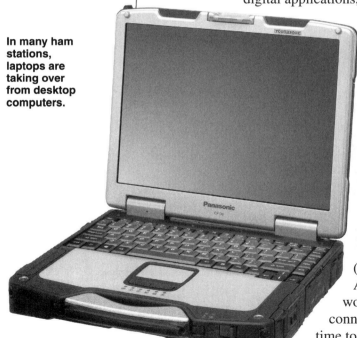

If you are considering a new or used computer (desktop or laptop), here are a few rule-of-thumb shopping specifications:

• The processor clock frequency ("speed") should be 1.5 gHz or better.

• The computer should have as much memory as possible, not just to run the ham applications smoothly, but the operating system as well. For *Windows Vista* or *7*, I'd recommend at least 4 GB of RAM; more is always better.

• Either a built-in wireless (Wi-Fi) modem or an Ethernet port. Although it isn't necessary for ham work, chances are you'll want to connect your station to the Internet from time to time. If so, you'll need a wireless modem or Ethernet port to do so.

• A CD-ROM drive for loading new software that is only available on CD (a less common need today, but don't sell yourself short).

The Rise of the Tablets

In 2011 the Gartner research group reported a sharp decline in desktop and even laptop computer sales. No doubt some of this decline was caused by the poor economy, but a large part was due to the meteoric rise of tablet computers.

The Apple iPad was among the first tablets out of the gate and it continues to dominate the market. If you own an iPad, or can travel to an Apple store to see one in the flesh, you'll instantly understand why. These thin, slick computers with gorgeously bright 10-inch screens are perhaps the perfect platforms for modern digital media. They can send and receive e-mail, display TV shows and movies, browse the Internet, act as e-book readers, become game platforms and do a great deal more. iPads are highly reliable and virtually immune to viruses. For hundreds of millions of people throughout the world, they are ideal go-anywhere computers.

Beyond the genius of its design, the iPad has risen to the consumer technology throne on the power of Apple's popular iTunes store, which offers more than 100,000 applications, known as *apps*, for just about every purpose imaginable.

Are the iPad and similar tablet computers the future of Amateur Radio?

Other tablet manufacturers are playing catch-up at this point. Amazon's tablet is gaining ground and others such as the Samsung Galaxy are carving out their market shares.

Some developers have already released apps for HF digital, primarily for the Apple iPad. Most notable among these are the *Multimode* and *I-PSK31* apps created by Luca Facchinetti, IW2NDH. Both of these apps allow you to send and receive RTTY and PSK31 —two of the most popular HF digital modes.

While tablets may indeed be the future of amateur HF digital, there remains the problem of interfacing tablets to transceivers. This can be a bit tricky and on the next page I offer some suggestions.

The Importance of Sound

With few exceptions, every computer you are likely to purchase today will include a either a dedicated sound card or a sound chipset. This feature is absolutely critical for HF digital operation because most of the modes you'll enjoy depend on sound

Tablet Connections

With the use of tablet computers becoming more widespread, we're seeing an increasing number of amateurs putting these devices to work in their stations.

When this book went to press, the most popular tablet by a wide margin was the Apple iPad. As we discuss in this chapter, there are already several iPad applications available for HF digital, specifically for RTTY and PSK31.

The main problem with using an iPad for HF digital is making the connections to the transceiver and keying the transceiver between transmit and receive. There are two ways of connecting iPads to the outside world:
- The microphone/headphone jack
- The docking port

If you purchase an Apple Camera Connection Kit and plug it into the docking port, you will suddenly have a USB connection to the iPad.

Unfortunately, this does not mean that you can plug in an HF digital USB interface. A USB interface draws more power than the iPad can provide, so the iPad will generate an error. An alternative is to use the Camera Connection Kit with a sound device designed for use with the iPad such as the Griffin iMic (**http://store.griffintechnology.com**).

The headphone/microphone jack is another option. This jack requires an unusual four-conductor plug. The connection diagram is shown in this sidebar. You can use this diagram along with a four-conductor plug to make your own transmit/receive audio adapter. The alternative is to purchase one premade from companies such as KV Connection at **www.kvconnection.com**. (Note that cables for iPhones will work for iPads as well.)

Now that you've solved the audio problem, what about keying the radio? The iPad provides no means whatsoever to attach any kind of keying device. The solution is to use the iPad's transmit audio to do the job. In the Appendix of this book you'll find a *QST* article that describes an HF digital interface designed by Skip Teller, KH6TY. His Digital VOX Sound Card Interface uses a Voice Operated Switch (VOX) circuit that keys the radio whenever transmit audio is sent from the iPad.

TigerTronics makes a device that operates on the same principle. Their SignaLink SL-1+ interface also uses a VOX keying circuit. The SL-1+ is available from TigerTronics at **www.tigertronics.com** at a cost of $80.

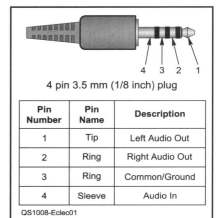

A diagram of the iPad headphone/microphone plug.

4 pin 3.5 mm (1/8 inch) plug

Pin Number	Pin Name	Description
1	Tip	Left Audio Out
2	Ring	Right Audio Out
3	Ring	Common/Ground
4	Sleeve	Audio In

QS1008-Eclec01

If you don't want to wire your own iPad audio adapter, you can buy one online.

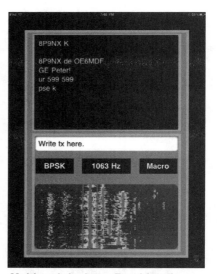

Multimode **by Luca Facchinetti, IW2NDH, receiving PSK31 on an Apple iPad.**

devices to act as radio *modems—mod*ulators/*dem*odulators.

The audio from your radio enters your computer via the sound device where it is converted (demodulated) to digital data for processing by your software. The results are words or images on your computer monitor. When you want to transmit, this same sound device takes the data from your software, such as the words you are typing, and converts it to shifting audio tones according to whatever mode you are using. This conversion is a form of modulation. The tones are then applied to your radio for transmission.

So how important are sound devices? Well, no sound device = no HF digital! (Well, there is one exception, which we'll discuss later.)

The simplest built-in sound devices are those found in laptops and tablets. They provide two ports to the outside world: microphone (audio input) and headphone (audio output). These are perfectly adequate for HF digital work. Desktop computers often have a similar arrangement, although the output port is usually labeled "speaker." A "line" input may also be included for stronger audio signals. In all cases these ports come in the form of ⅛-inch stereo jacks.

Some desktop computers offer sound cards that plug into the motherboard. These devices are more elaborate. Some sound cards can offer as many as 12 external connections. At the rear of your computer you may find LINE IN, MIC IN, LINE OUT, SPEAKER OUT, PCM OUT, PCM IN, JOYSTICK, FIREWIRE, S/PDIF, REAR CHANNELS or SURROUND jacks, just to name a few. For HF digital use, the important jacks are MIC or LINE IN and SPEAKER OUT.

Later in this chapter we'll discuss how to connect these sound devices to your radio, but one item needs to be briefly mentioned now: the *interface*. As you'll see later, the interface is yet another critical component because it is the link between the computer and the radio. The reason to mention it now is because the trend in interface technology has been to incorporate the sound device into the interface itself. At the time of this writing there are several interfaces by companies such as microHAM, West Mountain Radio and TigerTronics that feature their own built-in sound devices. These interfaces are extremely convenient because they work independently of whatever sound device you have in your computer. Just plug in their USB cables and

Laptops usually offer microphone and headphone audio jacks. These are fine for HF digital use.

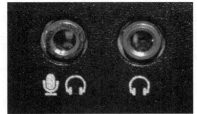

Rear panel motherboard audio ports in a typical desktop computer.

A deluxe sound card like this one installs inside a desktop PC. Note all the input and output ports.

you're good to go. It doesn't matter what kind of computer you are using; the interface will work with it. You also avoid a rat's nest of wiring between the computer and the radio.

Whether the sound device is in a plug-in card, a set of chips on the computer motherboard or a circuit inside an interface, one question often arises: Does the quality of the sound device affect your ability to operate?

This is one of the most common questions among HF digital operators. After all, the sound device is second only to the radio as the most critical link in the performance chain. A poor sound device will bury weak signals in noise of its own making and will potentially distort your transmit audio as well.

Before you dash out to purchase a costly high-end sound card, or obsess over the sound chipset in your interface or on your motherboard, ask yourself an important question: How do you intend to operate? If you have a modest station and intend to enjoy casual chats and a bit of DXing, save your money. An inexpensive sound card, or the sound chipset that is probably on your computer's motherboard or in your interface, is adequate for the task. There is little point in investing in a luxury sound device if you lack the radio or antennas to hear weak signals to begin with, or if they cannot hear you.

On the other hand, if you own the station hardware necessary to be competitive in digital DX hunting or contesting, a good sound card can give you an edge. This is particularly true if you are using a software defined radio. Sound card performance is critical for this application.

In 2007 the ARRL undertook an evaluation of 11 common sound card models. The study was performed by Jonathan Taylor, K1RFD, and the results were published in the "Product Review" section of the May 2007 issue of *QST*. As you'd expect, the high-end sound cards came out on top, but think carefully before you reach for your wallet. Don't buy more performance than you really need.

Software

When we're talking about putting together an HF digital station, a brief word about software is in order.

I say "brief" because it isn't practical to discuss software in detail within the pages of a printed book. Software evolves too rapidly for a book that has a useful lifetime measured in years. Any specific details would be obsolete almost before the book left the printing press. Instead, let's talk about software in broad strokes, beginning with operating systems.

As I mentioned earlier, most computers in Amateur Radio stations are running on some form of Microsoft *Windows*. As of 2011 quite a few amateurs were still using *Windows XP* or *Vista*, but the migration to *Windows 7* was well underway. By 2013 it is likely that most *Windows* users will be running *Windows 7*.

Apple Macintosh owners are obviously quite fond of the various *MacOS* incarnations, which have proven themselves to be efficient and reliable operating systems. The universe of Mac users is growing, but Amateur Radio software titles for the Mac are still few in number.

The *Linux* operating system in its various versions enjoys a loyal following among amateurs who like writing their own software and tinkering at the operating system level. Like the Mac, there is ham software available, but the offerings are somewhat sparse.

The three operating systems have their vocal advocates and I'd be a fool to champion one over another. All have their advantages and disadvantages to consider from an Amateur Radio point of view.

Windows

Pro: Sheer variety. Most Amateur Radio software is written for *Windows* so you have a rich selection of compatible software to choose from.

Con: Because of the widespread use of *Windows*, it is a favorite target for hackers. Anti-virus software is a must and this can significantly hamper the efficiency of *Windows* and cause other annoying issues. Also, *Windows* can be expensive if purchased and installed separately.

MacOS

Pro: Stability and performance. Highly intuitive and easy to use. Also, hackers have only rarely targeted *MacOS*.

Con: *MacOS* runs only on Apple Macintosh computers. It can be made to run on PCs, but it isn't an exercise for the fainthearted. Amateur Radio software for *MacOS* is limited.

Linux

Pro: Open source and free of charge. Depending on the version, can be quite efficient and powerful. Because there are so many versions in the field, hackers generally don't bother designing viruses for it. They prefer easier prey.

Con: Directory structure and commands may be very different compared to *Windows.* Amateur Radio software selection is very limited.

Beyond the operating systems, we have specific software for HF digital operating. In the beginning programs were designed for one specific mode,

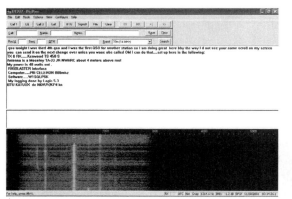

DigiPan software for PSK31.

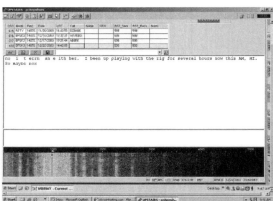

MixW, a popular multimode program for *Windows.*

MultiPSK for *Windows* offers a large number of HF digital modes.

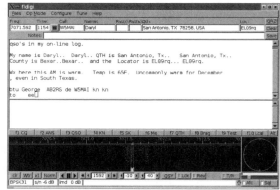

Fldigi is a multimode application that is available for both *Windows* and *Linux.*

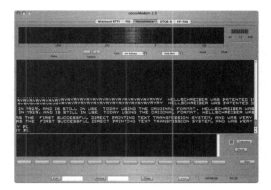

CocoaModem is another HF digital contender for Macs.

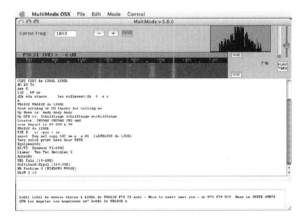

MultiMode by Black Cat Systems is HF digital software for Macs.

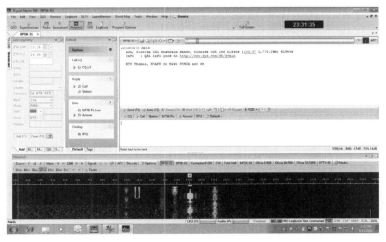

The *Digital Master 780* multimode HF digital program is part of the *Ham Radio Deluxe* package for *Windows.*

such as *DigiPan* for PSK31 (**www.digipan.net**) or *MMTTY* for RTTY (**http://hamsoft.ca/pages/mmtty.php**). While mode-specific software still exists, the trend has been strongly in favor of multimode software that can operate many different HF digital modes.

The most popular multimode applications are...
- *Windows*
    MixW **http://mixw.net**
    MultiPSK **http://f6cte.free.fr/index_anglais.htm**
    Fldigi **www.w1hkj.com/Fldigi.html**
- *MacOS*
    cocoaModem **www.w7ay.net/site/Applications/cocoaModem/**
    MultiMode **www.blackcatsystems.com/software/multimode.html**
- *Linux*
    Fldigi **www.w1hkj.com/Fldigi.html**

With the exception of *MultiMode* and *MixW*, all of the above are available for free downloading on the web. Until late 2011 *Ham Radio Deluxe*, another popular program for *Windows* was free for downloading at **www.ham-radio-deluxe.com/HRDv5.aspx,** but the software ownership has changed hands and it may no longer be free. Search Google for "Ham Radio Deluxe" for the latest information.

In later chapters we'll talk about features common to all these programs that make your operating more enjoyable.

The Interface—the Full Story

If you're like most amateurs, you already own most of the components we've discussed so far. You probably have an HF transceiver and, like 90% of most US households, you have a computer in residence. The one piece of hardware that you may not own is the one item that brings these components together to create an HF digital station: the *interface.*

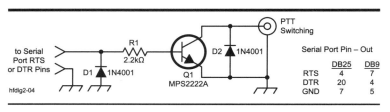

Figure 1.2 – The simplest way to key a radio using a computer is through a single-transistor circuit like this. The input connects to the serial cable coming from the computer (or from the USB/serial adapter). Either the RTS or DTR pins can be used, depending on what your software requires. The pin numbers for 25- and 9-pin plugs are shown.

At rock bottom an interface has only one job to do: to allow the computer to toggle the radio between transmit and receive. It achieves this by using a signal from the computer to switch on a transistor (see **Figure 1.2**). This transistor "conducts" and effectively brings the transceiver's *PTT* (Push to Talk) line to ground potential or very close to it. When the PTT line is grounded, the transceiver switches to transmit. When the signal from the computer disappears, the transistor no longer conducts and the PTT line is electrically elevated above ground. The result is that the transceiver returns to the receive mode.

The signal from the computer appears at a specific pin on a serial (COM) or USB port. Your HF digital software generates the signal when you click your mouse on TRANSMIT or some other button with a similar label.

If an interface can be so straightforward, couldn't you just build your own? Yes, you could. Many amateurs enjoy HF digital with simple interfaces like the one shown here. In addition to the switching circuit in Figure 1.2, they connect shielded audio cables between the computer and the radio to carry the transmit and receive audio signals. See **Figure 1.3**.

In Figure 1.3 you'll note that the audio lines include 1:1 isolation transformers. The reason for this is to avoid the dreaded *ground loop*. A ground loop results when current flows in conductors connecting two devices at different electrical potentials. In your HF station the conductors in question are usually the audio cables running between the radio and the computer.

A ground loop typically manifests itself as a hum that you'll hear in your receive audio, or that other stations will hear in your transmit audio. The hum can be so loud it will distort the received or transmitting signals, making digital communication impossible.

The isolation transformers effectively break the ground loop path while still allowing the audio signals to reach their destinations. These transformers are inexpensive, usually selling for less than $3 each from sources such as RadioShack.

To Roll or Not to Roll

There are good reasons to roll your own interface, the cost savings being chief among them. On the other hand, if you purchase an interface off the shelf you'll be able to

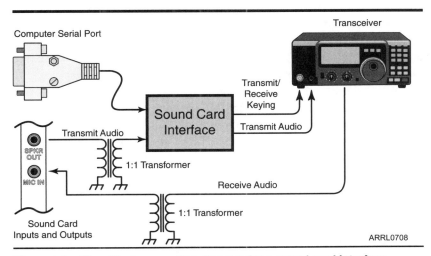

Figure 1.3 – Simplified connection diagram for a sound card interface.

benefit from enhanced design features, depending on how much you want to pay. The short list of useful features includes…

• **Independent Audio Level Controls.** These are knobs on the front panel of the interface that allow you to quickly raise or lower the transmit or receive audio levels. Many amateurs prefer to manage the audio levels in this fashion compared to doing it in software.

• **CW Keying.** Full-featured interfaces handle more than just HF digital. They can also use keying signals from the computer to send Morse code with a separate connection to the transceiver's CW key jack. This allows you to send CW from your keyboard rather than with a hand key, a useful feature for higher-speed CW exchanges during contests.

• **FSK Keying.** If you want to operate RTTY with the FSK function of your transceiver, assuming your rig offers such a function, the FSK keying feature translates keying signals from your computer into the MARK/SPACE data pulses necessary for FSK RTTY.

• **Microphone Input.** If you are making HF digital connections to your radio through the microphone jack rather than the rear panel accessory port, you'll need to unplug the interface cable whenever you want to use your microphone for a voice conversation. To make operating more convenient, some interfaces allow you to keep your microphone plugged into the interface at all times, switching between your microphone or computer as necessary.

• **Transceiver Control.** Remember that a basic interface does not allow your computer to truly control your radio, except in the sense that it can switch your radio between transmit and receive. Deluxe interfaces include the extra circuitry needed to allow full computer control of your transceiver. Sometimes referred to as *CAT* (Computer Aided Transceiver), this is a separate function that passes all the available controls from your radio to your computer. Depending on the type of transceiver you own and the software you are using, CAT allows you to change frequency, raise and lower power levels and much more. If your transceiver has the ability to connect directly to your computer through an RS-232 serial connection, USB cable or Ethernet port, you don't need the CAT feature. The CAT function is primarily intended for radios that use transistor-transistor (TTL) signals for control. Manufacturers sell their own CAT interfaces, but they tend to be expensive. An interface with CAT functionality brings everything together in one affordable box.

• **Built-in Sound Device.** As we discussed earlier, several interface designs include a built-in sound device. This is particularly handy in that it liberates the sound device in your computer for other functions. You can enjoy music on your computer, for example, without having to worry that you are meddling with the sound levels you've set up for HF digital operating. In addition, an interface

The microHAM USB III interface with a built-in sound device.

The MFJ-1275 interface.

The TigerTronics SignaLink USB interface also includes a built-in sound device.

The RigBlaster Advantage interface from West Mountain Radio has a built-in sound device.

A RigExpert interface.

with a built-in sound device greatly reduces the number of cables connecting the computer and radio. The audio signals, as well as transmit/receive keying functions, are all carried over a single USB cable; there are no connections to your computer's sound ports.

• **Pre-Made Cables.** Most commercial interface manufacturers include cables specifically wired for your radio free of charge, or offer them at an additional cost. This significantly reduces the hassle of wiring your HF digital station.

At the time of this writing, a basic off-the-shelf interface costs about $50; a multi-featured deluxe interface runs as high as $400. You'll need to shop among the manufacturers to find an interface that has the features you desire at a cost you are willing to pay. The most popular interface manufacturers include. . .

microHAM: **www.microham-usa.com**
MFJ: **www.mfjenterprises.com**
TigerTronics: **www.tigertronics.com**
West Mountain Radio: **www.
westmountainradio.com**
RigExpert: **www.
rigexpert.com**

Putting it all Together

If you've chosen an interface with a built-in sound device, assembling your station is relatively easy. You'll need a set of audio and PTT cables to connect the interface to your transceiver, either at the microphone and headphone jacks, or at the accessory jack. As I've already mentioned, you can purchase this cable from the interface manufacturer or make your own. The USB cable from the interface simply plugs into your computer.

The USB connection to your computer can be a little tricky in one respect, though. When you plug the USB cable into the computer for the first time, the computer may attempt to load and run a *driver* application so it can "talk" to your interface. This driver may already exist on your computer, or you may need to load it from a CD supplied by the manufacturer. Once the driver is loaded, the computer will recognize the interface every time you plug it in thereafter.

Even though the interface is connecting to your computer through a USB cable, it is depending on good old-fashioned serial communication just as though the connection had been made through a COM port. The interface accomplishes this by creating a *virtual COM port* in your computer. In other words, it uses software to emulate the function of a COM port.

Why is it important for you to know this? The answer is that your digital software will need to be configured so that it "knows" which COM port to use for PTT keying, CAT functions, etc. That means you'll need to know this as well!

In *Windows* it is a matter of going to the **Control Panel** and hunting down the **Device Manager** icon. Once you've started Device Manager, click on the **Port** section and you'll see all your computer ports listed in order. Look for a port labeled "USB Serial Port" or "Virtual COM Port" (see **Figure 1.4**). Next to it you'll see a COM number. Write this number down because you'll need to enter it when you're setting up your HF digital program and it asks for the "serial port" or "PTT port" (**Figure 1.5**).

When the computer recognizes the USB interface, it also recognizes the sound device within the interface. It will consider this device as another sound unit, just as though you had installed a second sound card inside the computer. (The computer doesn't know that this sound device is sitting in a box a few feet away and it doesn't care!) Again, this is important to understand because when you set up your software you may need to specify which sound device the software should use.

Figure 1.4 – *Windows* Device Manager showing the available ports. Note the "USB Serial Port" is labeled "COM 11."

Figure 1.5 – In this configuration screen from *Ham Radio Deluxe* you'll notice that COM 1 has been selected as the transmit/receive keying port.

Obviously you will need to select the sound device in your interface. Most software applications have drop-down menus that will list the available sound devices automatically. Don't expect to see your interface device listed by brand name. Instead, it may show up as "USB Sound," "USB Audio Codec" or something similar.

A 9-pin male serial port. These are rapidly disappearing from computers.

Non-USB Interfaces

If your interface doesn't have a USB connection, it likely uses a traditional serial connection instead. If your computer has a serial (COM) port, you need only attach a serial cable (typically a cable with 9-pin plugs at both ends) between the computer and the interface.

In most computers these serial ports fall in a range between COM 1 and COM 4. Plug your serial cable into an available port and use a bit of trial and error to find out which one you've selected. Start your HF digital program and go to the configuration menu. Enter a "1" into the COM port selection box and then click on the TRANSMIT button in your software. If your transceiver goes into transmit, congratulations—you've found the correct COM port. If not, try 2, 3, 4, etc.

But What if You Have a Serial Interface and Your Computer Doesn't Have Serial Ports?

Serial ports are going the way of the dinosaurs. You'd be hard pressed to find a desktop PC with serial ports today; in laptops they are nonexistent.

If your interface requires a serial cable, you have a problem. Fortunately, there are two solutions available.

(1) If you own a desktop computer that lacks serial ports, you can purchase a serial port card for less than $25. The card installs in the slots on the computer motherboard and offers one or two 9-pin serial port connections on the rear panel.

If you want to add traditional serial ports to your desktop computer, an

(2) If the thought of tinkering inside your desktop computer fills you with dread, or if

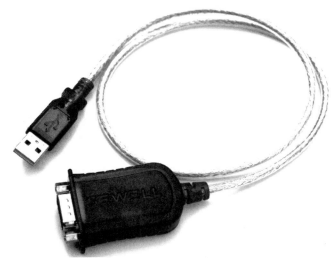

you own a laptop, you can use a USB-to-serial adapter cable. At one end of the cable you'll find a male USB plug; at the opposite end is a 9-pin (DB-9) male serial connector. When you plug the cable into your computer, it creates the "virtual COM port" described previously. These adapter cables are commonly available and cost less than $20.

A USB-to-serial adapter cable like this one makes it possible to attach serial interfaces to computers that only have USB ports.

Managing the Audio Connections

If you own an interface with a built-in sound device, you'll need to connect the transmit and receive audio cables between the interface and the transceiver, either at the microphone and headphone jacks, or at the accessory jack. If your interface is of the simple transmit/receive switching variety, the one or both sets of audio cables may have to go all the way back to the computer.

Even though you're using shielded audio cables, there is the potential for trouble when RF is in the air. This is especially true when your station antenna is close to your operating position. The audio cables can act like antennas themselves, picking up RF and wreaking havoc on your station. I've seen some instances where the computer shut down or reset whenever the transceiver was keyed. In other examples the RF energy mixed with the transmit audio and resulted in a horrendously distorted output.

If you suspect you have an RF interference problem in your HF digital station, you can diagnose it by reducing your output power and observing the results. If 100 W output gives you grief but 50 W is smooth as silk, you clearly have an RF interference issue.

Presumably you've kept your audio cables are short as possible. Stringing up 20-foot-long audio cables between the radio, the interface and the computer is just asking for trouble.

But if your audio cables are of reasonable length and you still suffer interference, it is time to buy some *toroid cores*. These are circular donuts made of a powdered iron and epoxy mixture. They come in various sizes and are rated for suppression at various frequencies. For HF applications, Type 61 toroids are

By wrapping cables through ferrite toroids, you can suppress or eliminate RF interference.

among the most effective. To suppress RF on an audio cable, wrap the cable through the toroid at least 10 times with evenly spaced turns.

With the right toroid in the right place, you can greatly reduce or eliminate RF interference. For severe cases you may need to place a toroid on every cable.

You'll often see used toroids for sale at hamfest fleamarkets, but don't buy a used toroid unless you know the type of material it contains. When in doubt, buy toroids new from manufacturers such as Amidon at **www.amidoncorp. com**. Avoid snap-on ferrite cores. While they are certainly easy to use, they are not as effective as toroids that you wind yourself.

Setting Up the Transceiver

Much of the advice that follows depends on what sort of transceiver you own and what kind of interface you are using to create your HF digital setup. When in doubt, always consult your transceiver and interface manuals.

If your interface is connecting to the radio through the transceiver accessory port, see if there is a function in the transceiver to adjust the accessory audio input and output levels. If it exists, this is a convenient way to establish "baseline" audio levels for the radio. Of course, you can also adjust audio levels at your computers and possibly at your interface (depending on the kind of interface you purchase). If it seems as though you aren't getting enough audio from the radio, or if it seems that you can't drive the radio to full output regardless of the computer or interface settings, check these transceiver settings as well.

Audio Overdrive

Speaking of driving rigs to full output, we need to discuss the danger of audio overdrive. Without question this is one of the most common issues among new HF digital operators.

Regardless of the type of digital mode you enjoy, there is an almost instinctual tendency among new operators to adjust their transmit audio levels while only watching their transceiver's RF output meter. They place their radios into the transmit mode and crank up the audio levels at their computers or interfaces until they see a satisfying 100 W RF output. At that point they assume they are finished and ready to take to the airwaves. This is known as "tuning for maximum smoke."

They could not be more mistaken!

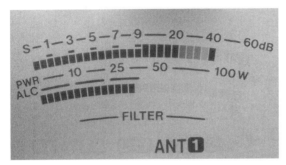

First of all, most HF digital modes do not require 100 W of power to make contacts. One of the benefits of HF digital, in fact, is that you can make contacts at surprisingly low power levels. For some digital modes such as JT65, 100 W output is considered obscene!

But most importantly, adjusting for full RF output ignores the fact that you may be grossly overdriving your radio to achieve your satisfaction. The result is often a wildly distorted signal that's not only difficult to decode, it splatters across the band, ruining every conversation in its wake. The ham who insists on generating a hideous signal for the sake of a few extra watts is known in traditional parlance as a LID. (No, it isn't a term of endearment.)

Rather than gazing at the transceiver meter as it displays your RF output, switch the meter to monitor *ALC* (Automatic Limiting Control) instead. All transceivers display ALC activity differently. The display may simply indicate the presence and amount of transmit audio limiting taking place. Other displays may include a "safe zone." If the ALC activity remains within the safe zone, your transmit audio levels are acceptable.

When you transmit a test signal, do not increase the audio level beyond the ALC safe zone, or beyond the point where ALC activity is excessive. When you see the needle or LEDs swing hard to the right, this is a warning that you are supplying way too much audio to the radio

Feeding too much audio to your transceiver can result in outrageously distorted transmit signals—like this one. Notice how the distortion products span the entire received bandwidth, causing grief to everyone.

In the case of this Kenwood TS-2000 transceiver meter, the ALC metering scale is on the bottom. As you can see, we're overdriving the radio with excessive audio, causing a substantial amount of ALC activity. Ideally, this meter should be reading zero.

In this ICOM transceiver meter, the ALC "safe zone" is indicated with a red bracket at the bottom. As long as the needle stays within the zone, you have a decent chance of transmitting a clean signal.

and that the ALC circuit is trying to rein you in.

The goal is to generate the desired RF output while keeping ALC activity to a minimum (or even zero), or while keeping the ALC meter in the safe zone. It is important to keep in mind that minimal ALC activity does *not* necessarily guarantee a clean signal. It does in many instances, but your best insurance is to ask for reports whenever you are in doubt. If someone reports that you are splattering, reduce the transmit audio level until they say your signal is clean. Note this setting so that you can return to it again easily.

Computer Sounds

When you begin listening to HF digital signals, don't be surprised if you occasionally hear beep, dings, chimes or a disembodied voice declaring, "You've got mail!"

Most of this interference isn't deliberate. It is caused by VOX-type interfaces that key transceivers whenever they detect audio – *any audio* – from the computers. I'm willing to bet the operators aren't even aware that their computers are guilty of this obnoxious behavior.

The solution, at least in *Windows*, is simple: Turn off "*Windows* sounds" before you get on the air. You can also do this manually by opening *Windows'* CONTROL PANEL and double-clicking the SOUND and AUDIO DEVICES icon. Click the Sound tab and under SOUND SCHEMES select NO SOUNDS. (Depending on the *Windows* version in question, the labels may differ.) Your fellow hams will thank you!

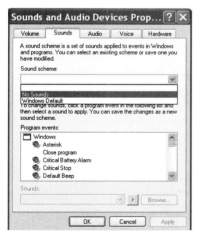

You can guard against transmitting odd *Windows* noises by turning off *Windows* sounds in Control Panel.

CAT Communication

If your software and interface support full transceiver control (CAT), you'll need to make sure that the data communication rates between the interface or computer and the transceiver are the same. Some clever pieces of CAT software will automatically analyze the data from the transceiver and quickly determine the data rate. For others, you'll have to enter a menu and specify the data rate.

In most CAT-capable radios you'll find a menu setting that will allow you to specify a data rate, often expressed as "baud." A rate of 9600 baud, for example, is common. You may need to access this menu to find out what setting the radio is currently using, or to change the radio's data rate to something the interface or software can handle.

Sampling Rates and
Windows XP

All sound devices operate by taking analog signals and chopping them into digital data. They do this by sampling the signal at a very fast rate, taking a kind of digital snapshot of the signal characteristics at a particular moment in time. The sampling rate is measured in Hertz (Hz).

Most HF digital programs rely on a sound device sampling rate of 11,025 Hz. However, some *Windows XP* computers fudge the sampling rate by a tiny amount. You wouldn't think this wouldn't cause a problem, and most of the time it doesn't, but there are other times when it becomes a serious issue.

If this error occurs on your computer, it can result in your transmit audio frequency being out of sync with the apparent receive frequency. When operating digital modes, this can manifest itself in several different ways. With PSK31, other stations may not respond when you answer their CQ, or they may respond but say that you are off frequency. It can also result in both you and the person that you're talking to "walking" across the band as each of you corrects your tuning after each transmission.

The good news is that this problem can be fixed by simply changing the sample rate in your HF digital program from 11,025 Hz to 12,000 Hz. Almost every program worth having has some means to do this; you may just have to go digging through the menus.

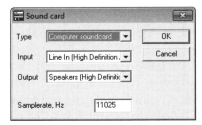

Changing the sample rate in *DigiPan*.

When an Interface Isn't an Interface—PACTOR

Throughout this chapter we've held fast to an overarching assumption: that the mode you will choose to operate will use software that transmits and receives through a sound device. The sound device can be inside the computer or within the interface. Either way, it is at the heart of your ability to communicate.

There is one exception to this assumption: *PACTOR*.

PACTOR is a digital mode invented by two German amateurs in 1991 and it has been evolving ever since. PACTOR is unique because it is capable of 100% error-free communication at global distances on the HF bands. This makes PACTOR the hands-down favorite when messages must get through error-free. That's why PACTOR is the most widely used digital mode among those who participate on the HF side of the Winlink network. We'll discuss Winlink later in this book.

The SCS PTC-IIusb is a PACTOR controller that offers several other modes as well.

PACTOR works its magic by sending data in small chunks or *frames*. The receiving station analyzes the data and keeps whatever has arrived without errors. It transmits a short burst back to the originating station and tells it to repeat the frame. With luck, the next retransmission of the frame will arrive with most of the data intact, or with at least enough intact data that the receiving station can mix and match between the "new" and "old" frames to come up with a complete 100% error-free frame.

All of this back-and-forth transmitting makes a characteristic *chirp-chirp-chirp* sound on the air. It also requires a fast-switching transceiver and a lot of computational horsepower. To this date no one has written a computer program that is capable of doing PACTOR with sound devices. The PACTOR timing parameters are too strict and computer operating system timing tends to be too loose.

So, the manufacturers have packaged dedicated microprocessor circuitry into a stand-alone box. This box, often called a *controller*, has the speed and efficiency to meet the requirements for PACTOR. It can do other modes as well, such as PSK31 and RTTY. Unlike a desktop or laptop computer, a controller doesn't labor under the burden of an operating system and it isn't required to fulfill a broad range of simultaneous tasks.

Sounds like a simple, elegant solution, doesn't it? It's reasonable to ask why anyone would bother using sound devices to enjoy digital operating when a stand-alone controller can do the job.

The answer is cost. The newest PACTOR controllers retail at about $1000+. And don't forget that you still need a computer and software to communicate with the controller.

With most sound-device software selling at a low cost or even zero cost, and with even high-end interfaces selling at less than $400, the cost to become active in sound-device-based HF digital—assuming you already own a computer—is substantially less than the cost of a PACTOR controller.

If you really want PACTOR capability, you must purchase a controller—no two ways about it. On the Winlink e-mail network you may be able to get away with an older PACTOR I controller that you'll find selling for a few hundred dollars, but to step up to modern PACTOR II you will most definitely need the $1000+ units manufactured by SCS Corporation on the Web at **www.scs-ptc. com**.

The SCS model PTCIIusb is a good choice because it offers a wide range of features in a controller that plugs directly into your computer's USB port. You can purchase a set of cables to connect the PTCIIusb to your transceiver (an additional $50). In addition to PACTOR II, this controller will do PSK31, RTTY, CW, Slow Scan TV, packet radio and more.

Keep in mind that to be successful with PACTOR your transceiver must be able to switch between transmit and receive in less than 20 milliseconds. Many rigs can do this, but if you are in doubt, check the *QST* magazine Product Reviews. If you are a member of the ARRL, you'll have full access to the Product Review archives online at **www.arrl.org/product-review**.

Chapter Two

PSK31

The Short Scoop
PSK31 is the most widely used HF digital operating mode today. Like CW, its chief advantage is that its signals are very narrow (energy and spectrum efficient) and it can be decoded under very poor conditions.

When sound card technology took over the amateur HF digital world in the year 2000, PSK31 was the vanguard mode. It was the brainchild of Peter Martinez, G3PLX. He wanted to create a mode that was easy to use yet extremely robust in terms of weak-signal performance. Another important criterion was bandwidth. The HF digital subbands are narrow and tend to become crowded in a hurry (particularly during contests). Peter wanted to design a mode that would do all of its tricks within a very narrow bandwidth.

In early 1999 Peter unveiled the first PSK31 software for *Windows*. The program required only a common 16-bit sound card functioning as the analog-to-digital converter (and vice versa). With the large number of hams owning PCs equipped with soundcards, the popularity of PSK31 exploded.

Today PSK31 is the undisputed king of HF digital. Just go to any of the PSK31 "watering holes" (see **Table 2.1**) such as 14.070 MHz and you'll hear its distinctive warbling signal at just about any time of the day or night.

Table 2.1
Popular PSK31 Frequencies

3580 kHz
7070 kHz
10140 kHz
14070 kHz
21070 kHz
28120 kHz

What is PSK31?

First, let's dissect the name. The "PSK" stands for Phase Shift Keying, the modulation method that is used to generate the signal; "31" is the bit rate. Technically speaking, the bit rate is really 31.25, but "PSK31.25" isn't nearly as catchy.

Think of Morse code for a moment. It is a simple binary code expressed by short signal pulses (dits) and longer signal pulses (dahs). By combining strings of dits and dahs, we can communicate the entire English alphabet along with numbers and punctuation. Morse uses gaps of specific lengths to separate individual characters and words. Even beginners quickly learn to recognize these gaps; they don't need special signals to tell them that one character or word has ended and another is about to begin.

PSK31 transmits its binary code in an interesting way. Instead of keying the signal on and off, PSK31 uses Digital Signal Processing to create an audio signal that shifts its *phase angle* 180° in sync with the 31.25 bit-per-second data stream. In Peter's scheme, a 0 bit in the data stream generates a phase shift, but a 1 does not. So, PSK31 software uses two phase angles (0° or "in phase" and 180° out of phase) to communicate 1s and 0s. **See Figure 2.1**. Because two phase angles are used to accomplish this, it is usually referred to as *binary* phase-shift keying, or BPSK (most people use the term "PSK31" as a kind of catch-all label).

If you apply a BPSK audio signal to an SSB transceiver, you end up with BPSK modulated RF. At this data rate the resulting PSK31 RF signal is only 31.25 Hz wide, which is actually narrower than the average CW signal.

Concentrating your RF into a narrow bandwidth does wonders for reception, as any CW operator will tell you. But when you're trying to receive a BPSK-modulated signal, it is easier to recognize the phase transitions—even when they are deep in the noise—if your computer knows when to expect them. To accomplish this, the receiving station must

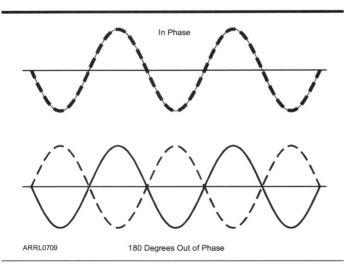

ARRL0709 180 Degrees Out of Phase

Figure 2.1 – A BPSK signal communicates binary information (0s and 1s) by shifting the signal phase 180 degrees. This phase shifting is what gives PSK31 its warbling sound.

synchronize with the transmitting station. Once they are in sync, the software at the receiving station "knows" when to look for data in the receiver's audio output. Every PSK31 transmission begins with a short "idle" string of 0s. This allows the receive software to get into sync right away. In PSK31 the phase transitions are also mathematically predictable, so much so that the PSK31 software can quickly synchronize itself automatically when you tune in during the middle of a transmission, or after you momentarily lose the signal.

The combination of narrow bandwidth, an efficient DSP algorithm and synchronized sampling creates a mode that can be received at very low signal levels.

Its terrific performance notwithstanding, PSK31 will not always provide 100% copy; it is as vulnerable to interference as any digital mode. And there are times, during a geomagnetic storm, for example, when ionospheric propagation will exhibit poor frequency stability. When you are trying to receive a narrow-bandwidth, phase-shifting signal, frequency stability is very important.

The Flavors of PSK31

Many people urged Peter to add some form of error correction to PSK31, but he initially resisted the idea because most error-correction schemes rely on transmitting redundant data bits. Adding more bits while still maintaining the desired throughput required a doubling of the data rate. If you double the BPSK data rate, the bandwidth doubles. As the bandwidth increases, the signal-to-noise ratio deteriorates and you get more errors. It's a sticky digital dilemma. How do you expand the information capacity of a BPSK channel without significantly increasing its bandwidth?

Peter finally found the answer by adding a second BPSK carrier at the transmitter with a 90° phase difference and a second demodulator at the receiver. The result is a signal that consists of four waveforms that differ in phase, each 90 degrees apart. Peter calls this quadrature polarity reversed keying, but it is better known as *quaternary phase-shift keying* or QPSK.

Operating PSK31 in the QPSK mode will give you 100% copy under most conditions, but there is a catch. Tuning is twice as critical with QPSK as it is with BPSK. You have to tune the receive signal within an accuracy of less than 4 Hz for the decoder to detect the phase shifts and do its job. Obviously, both stations must be using very stable transceivers.

There are other forms of PSK31 that use higher data rates. Of course, the higher the speed, the wider the signal. As you prowl the bands you may occasionally hear the burr of PSK63, 125 or even 250. Like QSPK, however, these modes are not common. Almost every warble you'll hear on the air will be BPSK31.

Does the Sideband Matter?

With BPSK, the most popular PSK31 mode by far, the answer is "no." You can have your transceiver in upper or lower sideband and work any signals that appear in your software display.

QPSK is another matter. Sideband selection *is* critical for QPSK. Most QPSK operators choose upper sideband, although lower sideband would work just as well. The point is that both stations must be using the *same* sideband, whether it's upper or lower.

Software Evolution and the Panoramic Solution

One of the early bugaboos of PSK31 had to do with tuning. Most PSK31 programs required you to tune your radio carefully, preferably in 1-Hz increments. In the case of the original G3PLX software, for example, the narrow PSK31 signal would appear as a white trace on a thin waterfall display. Your goal was to bring the white trace directly into the center of the display, then tweak a bit more until the phase indicator in the circle above the waterfall was more-or-less vertical (or in the shape of a flashing cross if you were tuning at QPSK signal). Regardless of the software, PSK31 tuning required practice. You had to learn to recognize the sight and sound of your target signal. With the weak warbling of PSK31, that wasn't always easy to do. And if your radio didn't tune in 1-Hz increments, the receiving task became even more difficult.

Nick Fedoseev, UT2UZ and Skip Teller, KH6TY, designed a solution and called it *DigiPan*. The "pan" in *DigiPan* stands for "panoramic." With *DigiPan* the idea is to eliminate tedious tuning by detecting and displaying not just one signal, but *entire groups of signals*.

If you are operating your transceiver in SSB without using narrow IF or audio-frequency filtering, the bandwidth of the receive audio that you're dumping to your sound device ranges from about 100 Hz to 3000 Hz. As you might imagine, with a bandwidth of only 31 Hz, many PSK31 signals can squeeze into that 2900 Hz chunk of spectrum. Panoramic software acts like an audio spectrum analyzer, continuously sweeping through the received audio spectrum and showing you the results in a large *waterfall display* that continuously scrolls from top to bottom. What you see on your monitor are vertical lines of various colors that indicate every signal that the software can detect. Bright lines represent strong signals while faint lines indicate weaker signals.

Other software writers jumped on the panoramic bandwagon and soon nearly every PSK31 program used waterfall displays of one type or another. The beauty

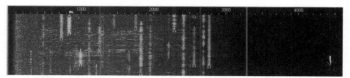

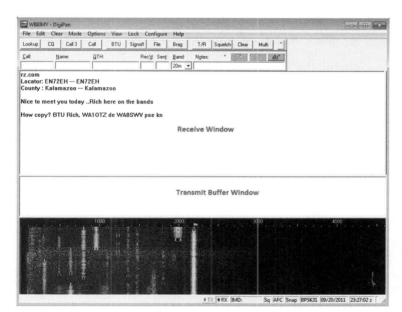

DigiPan and many other HF digital programs display signals in what is known as a *waterfall*. In this example, the white lines represent individual PSK31 signals. To decode a signal, all you have to do is click your mouse cursor on one of the lines.

of panoramic reception is that you do not have to tune your radio to monitor any of the signals you see in the waterfall. You simply move your mouse cursor to the signal of your choice and click. A cursor appears on the trace and the software begins displaying text. You can hop from one signal to another in less than a second merely by clicking your mouse.

Getting on the Air – Step by Step

Let's take a step-by-step approach to your first PSK31 contact. To keep things simple, we'll use *DigiPan* software for *Windows* as our example. *DigiPan* is free to download and easy to understand. It runs well on *Windows XP*, *Vista* and *7*.

If you are a Mac owner, I'd recommend you start with W7AY's *cocoaModem*, which is available free of charge at **www.w7ay.net/site/ Applications/cocoaModem/**. *Linux* users may want to start with *Fldigi* at **www. w1hkj.com/download. html**. Both are multimode programs that offer PSK31, but many of the same ideas you'll learn about *DigiPan* apply to multimode software as well, only the labels and menus may change.

If you'd like to follow along with your computer, you can download *DigiPan* from **www.digipan.net**, or from the special *Get on the Air with HF Digital* page on the ARRL website at **www. arrl.org/arrl-store/HF- Digital**. Note that *DigiPan* was written before the advent of *Windows Vista* or *7*. As a result, the *DigiPan*

DigiPan is an easy-to-use *Windows* program for PSK31. In this example, we've added labels to indicate the RECEIVE window and the TRANSMIT BUFFER window.

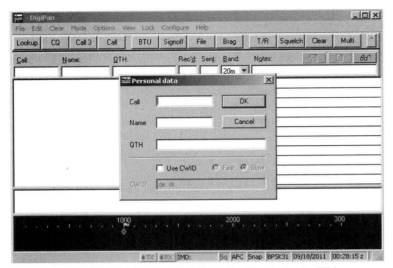

When you run *DigiPan* for the first time, it will ask you to fill in your call sign, name and location (QTH). Go ahead and do so. You can leave the CW ID box unchecked.

"help" files may not appear when you click on the HELP menu. It is a programming quirk of *Vista* and *7*. Even so, *DigiPan* is so intuitive you probably won't need to resort to HELP as you explore the features that extend beyond the basics we'll cover here.

We'll begin by assuming that you've installed *DigiPan* on your hard drive. When you run *DigiPan* for the first time, it will ask you to fill in your call sign, name and location (QTH). Go ahead and do so. You can leave the CW ID box unchecked.

Throughout this exercise I'll also assume that you have your interface hooked up between your computer and your HF transceiver.

Configuring the Sound Card and Serial Port

Before we get started, we need to discuss a potentially vexing oddity that can occur with some sound device *driver* applications. A driver is a small piece of software that is dedicated to a specific task, typically "talking" to various devices inside or outside the computer. A driver acts as the liaison between the software and the device the software is attempting to use. For example, a printer driver is the liaison between a word processing program and the printer hardware.

In an effort to provide more user convenience, some sound device drivers attempt to detect when something (such as a microphone) has been plugged into the audio ports. In theory, the drivers should then automatically route the audio signals to the software accordingly.

This is a fine concept, but it can cause headaches for HF digital software users. The pain occurs when you start your HF digital application (such as *DigiPan*) and instantly receive an error message telling you that the "Sound card is in use or does not exist." This happens because the driver isn't allowing your HF digital software to gain access to the audio data from the sound device. Fortunately, the solution is simple: plug in the audio cables from the interface, or

If you see this error message when you start *DigiPan* or any other sound-device-based program, it means that the sound device driver isn't allowing your HF digital software to gain access to the audio data from the sound device. This is often caused by cables not being plugged into the computer, or because some other program is making exclusive use of the sound device.

DigiPan's CONFIGURE menu allows you to set up the program to your liking.

Once you have *DigiPan* up and running, click CONFIGURE and then click SOUND CARD. In the little window that appears you want to make sure that the sound device "Type" is correct (either your internal sound card or external sound device) and that *DigiPan* is using the correct inputs and outputs.

plug in the USB cable if the interface contains its own sound device. The driver will be satisfied and will allow your HF digital software to access the sound device.

Once you have *DigiPan* up and running, click CONFIGURE and then click SOUND CARD. In the little window that appears you want to make sure that the sound device **Type** is correct (either your internal sound card or external sound device) and that *DigiPan* is using the correct audio inputs and outputs. For example, if you are connecting the transceiver's receive audio to the LINE INPUT of your computer sound card, make sure that the LINE INPUT has been selected. Don't worry about the Sample rate window.

In the *DigiPan* CONFIGURE menu, click SERIAL PORT. Here is where you'll select the computer port you will use for Push to Talk (PTT) to key the

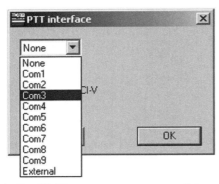

In the *DigiPan* CONFIGURE menu, click SERIAL PORT. Here is where you'll select the computer port you will use for Push to Talk (PTT) to key the transceiver. Highlight the COM port your interface is using.

transceiver. Highlight the COM port your interface is using. Remember that if you have an interface with a USB cable, or if you are using a USB-to-serial adapter, you need to select the "virtual COM port" number. Re-read Chapter 1 if you've forgotten.

Also, check the boxes labeled RTS AS PTT and DTR AS PTT. By doing so you're telling *DigiPan* to send keying pulses on both the Request to Send (RTS) and Data Terminal Ready (DTR) lines on the COM port. Your interface will be looking for the PTT keying signal from one of these lines; by checking both boxes you'll cover all your bases, so to speak.

There are many other features you can configure in *DigiPan* to customize it to your liking, but these will be sufficient to get you started.

Receiving PSK31

Turn on your transceiver, select Upper Sideband (USB) and tune to one of the PSK31 frequencies shown in Table 2.1. During daylight hours, 14.070 MHz is a good choice. If propagation conditions are favorable, you should hear the warbling sounds of PSK31 signals.

Look at the waterfall display (the bottom window) in *DigiPan*. If everything is working properly, you should see a screen similar to **Figure 2.2**. Each line represents a signal.

If your waterfall display is showing only faint blue lines or nothing at all, this means that not enough receive audio is getting to the sound device in your interface or computer. If the

Figure 2.2 – If *DigiPan* is running properly, you should see lines in the waterfall display that represent PSK31 signals. Click your mouse cursor on one of the lines and watch what happens.

The Sound panel in *Windows 7*.

When you click on the RECORDING tab in the Sound panel, you'll see a list of all the audio inputs and devices that *Windows* "knows about." The audio input or external sound device should be highlighted. It may also include a checkmark.

With the Recording device highlighted, click on the PROPERTIES button and you'll be presented with yet another window with more detail about the device in question. Click the LEVELS tab to adjust the audio input levels.

waterfall window is utterly black, the first things to check are your audio cables. Make sure they are plugged into the correct ports.

If you are using an interface that has receive and transmit audio controls on the front panel, make sure the receive audio control is turned up. If your interface lacks audio controls, it is time to look at the sound controls in your computer.

You can access the sound input level controls in *DigiPan* by clicking on the CONFIGURE menu, followed by WATERFALL DRIVE. However, this function only works if your computer is running *Windows XP*. If you are using *Windows Vista* or *Windows 7*, you'll need to jump through several hoops. From your *Windows* Start menu, open the CONTROL PANEL and double click on the SOUND icon. In the window that opens next, click the RECORDING tab.

Now you'll be presented with a list of all

the audio inputs and devices that *Windows* "knows about." The audio input or external sound device should be highlighted. It may also include a checkmark. If the correct input or device is not highlighted, click on it now.

With the device highlighted, click on the PROPERTIES button and you'll be presented with yet another window. Click the LEVELS tab and you'll finally see the slider control that will allow you to increase the audio input level. If it is at minimum, turn it up (slide it up or to the right).

If all is well, you should now see blue or yellow lines moving from top to bottom in the waterfall window. If the waterfall window is filled with a yellow haze and the signal lines are mostly red, you have too much audio. Reduce the receive audio level until only the strongest signals appear as yellow lines.

Examine the waterfall window closely. Do you see the numbers 1000, 2000 and 3000 along the top of the window? These are audio frequencies in Hertz (Hz). You may also notice that the blue haze in the waterfall is somewhat brighter at the far left and that it fades to black at the far right, probably just beyond the 3000 Hz marker.

What you are seeing is the entire audio bandwidth of your radio's receiver, ranging from 100 to about 3000 Hz. The waterfall haze is brighter at the lower end because the audio response of your radio is stronger there. If you own a transceiver that allows you to adjust receiver audio equalization, you can tweak it for a more uniform response from low to high frequencies, although this isn't really necessary for using *DigiPan*.

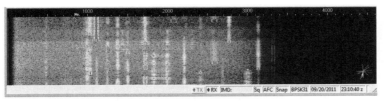

Too much input audio will fill the *DigiPan* waterfall with a bright haze.

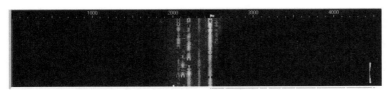

Switch in a narrow transceiver IF filter (such as a CW filter) while watching the waterfall display. You'll see the waterfall haze suddenly become much narrower according to the width of the filter. You'll also notice that all the other signals outside the filter bandwidth have vanished. This is a powerful illustration of how beneficial a narrow IF filter can be.

If you want to try something educational, switch in a narrow transceiver IF filter (such as a CW filter) while watching the waterfall display. You may be able to do this by simply switching your radio to the CW mode. You'll see the waterfall haze suddenly become much narrower according to the width of the filter. You'll also notice that all the other signals outside the filter bandwidth have vanished. This is a powerful illustration of how beneficial a narrow IF filter can be.

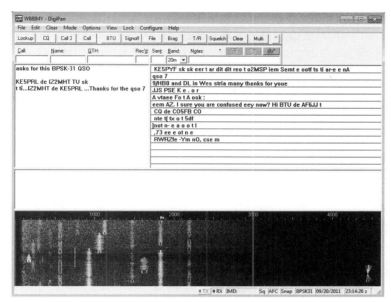

DigiPan offers a unique multichannel view that you can activate in the VIEW menu. When this feature is active, *DigiPan* will simultaneously decode every signal in the waterfall!

Switch back to USB and click your mouse cursor on one of the signal lines. You should be rewarded with text flowing across the top window. To tune to a different signal, all you have to do is click your mouse cursor on another line.

It is important to emphasize here that you are *not* tuning your radio as you hop from one signal to the next. Your radio's VFO display remains fixed at 14.070 MHz. That's the magic of panoramic reception within an audio bandwidth. Because a panoramic display frees you from the need to tune your radio, it is probably a good idea to lock your VFO if your radio provides that option. This will prevent you from accidentally bumping the knob and sifting frequency.

If you have several visible signals in the waterfall window, here is something fun to try. *DigiPan* offers a unique *multichannel* view that you can activate in the VIEW menu. When this feature is active, *DigiPan* will attempt to simultaneously decode every signal in the waterfall! This makes for a fascinating, if somewhat chaotic, display.

Using the waterfall display you may often see (and copy) PSK31 signals that you cannot otherwise hear. It is not at all uncommon to see several strong signals (the audible ones) interspersed with wispy blue ghosts of very weak "silent" signals. Click on a few of these ghosts and you may be rewarded with text (not error-free, but good enough to understand what is being discussed).

In the lower right corner of the waterfall window you'll notice green or multicolored lines that seem to move rapidly. This is the phase display. When you click on a strong BPSK signal, you'll see the lines turning green and aligning vertically (more or less). This is an indicator of how well *DigiPan* is decoding the binary phase shifts, so you can consider the display as an indicator of signal quality. With a perfect BPSK signal the green lines will be perfectly

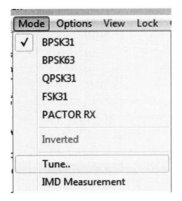

vertical. Of course, perfect conditions are uncommon so you'll usually see the green lines flickering all over the place. If the signal is terrible or very weak, the lines will be anything but vertical!

You don't need to pay attention to the phase display during normal operation, but it can occasionally tell you some interesting things. As an example, let's say that you've clicked your mouse cursor on what seems to be a strong signal, but the resulting print is garbled at best. Glance at the phase display. If the lines are flickering chaotically around the circle, this means that while the signal may be strong, the phase relationships are badly disrupted. This can occur when receiving signals that have come to you from over the Earth's poles and are suffering from "polar flutter." You'll see the same thing when receiving PSK31 from nearby stations whose signals are arriving at your antenna along several different pathways and partially canceling each other out ("multipath distortion").

Time to Tune

Now that you've spent some time decoding PSK31 conversations, no doubt you'll be eager to transmit. Look at the waterfall display and try to find an open spot between the signal lines. Click your mouse cursor on one of those empty spaces. You've just chosen your transmit frequency.

When you transmit, the PSK31 software generates a tone that corresponds to the frequency "position" in the waterfall where you clicked your mouse cursor. (Again, see the audio frequency scale along the top of the *DigiPan* waterfall display.) When that tone is applied to your radio, it creates an RF signal on the correct frequency – a certain number of Hertz above or below the suppressed carrier frequency. In the CONFIGURATION window you can set up *DigiPan* to display either the tone frequencies or the corresponding RF frequencies in the waterfall.

It is worthwhile to note that the frequency your transceiver displays in the SSB mode is the *suppressed carrier frequency*. If you've selected upper sideband on your radio (USB), your receive audio range is everything from the suppressed carrier frequency to about 2 or 3 kHz above it. That's what you are seeing in the waterfall display. *DigiPan* is continuously sweeping through this range and displaying the results. If you select lower sideband (LSB), your receiver range extends 2 or 3 kHz *below* the suppressed carrier frequency. We'll discuss this in more detail later.

Let's check your audio and RF output levels before we do anything else.

Switch your transceiver's meter to ALC. Now click the MODE button along the top of the *DigiPan* window and select TUNE.

You transceiver should jump to the transmit mode and stay there. Using your interface controls or your audio output control in *Windows* (this may appear as a tiny speaker icon in the lower right corner of the *Windows* desktop), increase the audio level until you *just* see ALC activity, then reduce the level until ALC activity ceases. If your transceiver meter uses an ALC "zone" instead, increase the audio level until the needle moves to the higher end of the zone.

Now quickly switch the meter to display RF output power. You should see full RF output according to the power level you've set on transceiver: 100 W, 50W or whatever. Click OK in the TUNE window to unkey your transceiver.

It is possible that you may see less RF output than you anticipated. This can be caused by the way the ALC function works in your particular radio; even minimal ALC activity may result in reduced output. Another possibility involves the *transmit audio response* of your radio. A brief explanation is in order.

Transmit Audio Response

Let's say you've chosen a transmit "location" in the *DigiPan* waterfall display that corresponds to an audio tone of 1000 Hz. When you transmit, your sound device will send a 1000 Hz tone to your radio. Since you are operating in Upper Sideband (USB), your radio will generate an RF signal 1000 Hz above the suppressed carrier frequency shown on the transceiver's display. If the display reads 14.070 MHz, you will be transmitting at 14.071 MHz (14.070 + 1000 = 14.071 MHz). See **Figure 2.3**.

If you select a point in the waterfall that corresponds to 300 Hz, your radio will transmit at 14.0703 MHz (14.070 + 300 = 14.0703 MHz).

So far so good.

You'd think that your RF output level would be the same

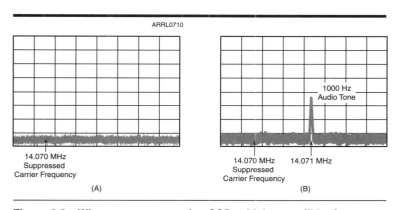

ARRL0710

(A)

1000 Hz
Audio Tone

14.070 MHz
Suppressed
Carrier Frequency

14.070 MHz   14.071 MHz
Suppressed
Carrier Frequency

(B)

Figure 2.3 – When you are operating SSB, which you will be for most digital modes, the frequency shown in the transceiver display is the *suppressed carrier frequency*. If you key your transceiver but don't apply audio, nothing happens. No RF is generated (A). However, if you key your transceiver and apply a 1000 Hz tone, the radio will generate an RF signal 1000 Hz above the suppressed carrier frequency. In this PSK31 example, you may think you are transmitting at 14.070 MHz, but in reality you are transmitting at 14.071 MHz.

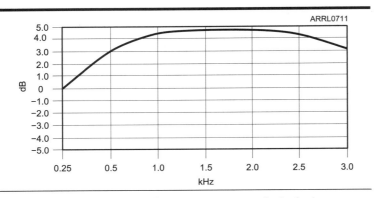

Figure 2.4 – The transmit audio response curve of a typical transceiver. Notice how the curve drops at the high and low ends. If you apply a 300 Hz tone to the radio, the resulting RF output will be lower than it would be if you applied a 1000 Hz tone.

regardless of whether you sent a 1000 Hz or 300 Hz tone to the radio, but you'd be wrong. The transceiver has a *response curve* that limits the degree to which the various audio frequencies are amplified before they are converted to RF. It applies to audio coming in at the accessory port or microphone jack. This response curve can differ from one brand of radio to another. In some radios it can be changed through a feature often referred to as "transmit audio equalization."

Generally speaking, the response tends to roll off at the high and low ends of the audio range. See **Figure 2.4**. What this means to you is that the strength of an audio tone at the low or high ends of the curve will be reduced within the transceiver before it becomes RF. The result will be reduced RF output even though the audio level from your sound device hasn't changed.

What can you do? There are a couple of options …

•Adjust your sound device output level to compensate.

•Find an open transmit frequency that is closer to the middle portion of the waterfall (around 1000 Hz).

If you encounter this behavior when you find an interesting station whose signal is at the extreme edge of the waterfall (high or low), another approach is to gently adjust your transceiver VFO while watching the waterfall and "move" the signal closer to the middle of the display.

Of course, the final option is to simply ignore the issue. PSK31 doesn't require much RF to work successfully, so reduced output probably won't make that much difference.

The T/R button on the *DigiPan* toolbar toggles your radio between receive and transmit.

Calling CQ

There are two ways to call CQ in *DigiPan*. One method, the one we'll discuss here, is best described as the "manual method." There is a more automated approach that we will address later.

For now, click your mouse cursor in the middle window of *DigiPan*. This is your *transmit buffer*. Everything you type in this window will be sent when it is time to transmit. Type the following, substituting your call sign for mine, of course…

CQ CQ CQ CQ WB8IMY WB8IMY WB8IMY CQ CQ CQ CQ WB8IMY WB8IMY WB8IMY CQ K

Now click the T/R button in the *DigiPan* tool bar. Your radio should pop into the transmit mode and begin sending your CQ message. You'll see your wavering RF output and you will also see the transmitted text in the upper receive window. Watch this text closely. When you see the "**K**" appear ("over" in Morse code), quickly click the T/R button again to place your radio back in the receive mode.

Now we wait! With luck someone will respond …

WB8IMY WB8IMY DE WJ1B WJ1B WJ1B K

From this point the conversation proceeds in turns. You type your reply in the transmit buffer and click T/R to send the text, and then wait for his response.

WJ1B DE WB8IMY... Thank you for the reply. Name is Steve Steve and I am in Wallingford Wallingford, CT. Your RST is 589. How copy? WJ1B DE WB8IMY K

DigiPan, like most HF digital programs, allows you to type in the transmit buffer while you are receiving. This is a *very* handy function – especially if you are a slow typist. As the other station transmits you can read his text and type comments in your transmit buffer as you go. When it is your turn, just click on the T/R button and everything you've typed will be transmitted. At his end you'll look like the world's fastest, smoothest typist!

Macros

So far we've discussed the manual method of sending PSK31 text with *DigiPan*. Now it is time to introduce a bit of automation with *macros*.

The word "macro" is from the Greek μακρό for "big" or "far." In the digital world a macro is a single instruction that causes an entire set of actions (or instructions) to take place. For HF digital software, a macro is used to send preformatted "canned" text.

Macros are commonly used during contests when you are tasked with sending the same text repeatedly. DX stations also use macros to send exchanges automatically when they need to work as many stations as possible, as quickly

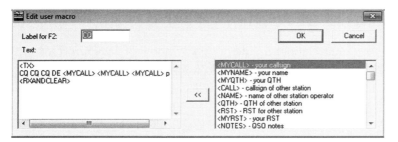

Figure 2.5 – *Right click* on the CQ button. A window will open to show you the preformatted macro. The text that appears inside the "less than" and "greater than" symbols (< and >) is interpreted by *DigiPan* as a command, not as text to be sent.

as possible. Hams in other countries will often rely on macros when they are not fluent in English. They prepare several text messages as macros so that they can send understandable English quickly and accurately.

A macro can be as simple or complex as you desire. *DigiPan* comes with several macros preformatted. These appear as buttons near the top of the window.

With your mouse, *right click* on the CQ button. A window will open to show you the preformatted macro (see **Figure 2.5**). The text that appears inside the "less than" and "greater than" symbols (< and >) is interpreted by *DigiPan* as a command, not as text to be sent.

<TX> CQ CQ CQ DE <MYCALL> <MYCALL> <MYCALL> pse K <RXANDCLEAR>

Strange as it may look, this is a straightforward macro. Let's break it down …

<TX>   This is a command to place the radio in the transmit mode.

CQ CQ CQ DE   This is plain text to be sent.

<MYCALL>   This is a commend to insert your call sign. It repeats three times.

Pse K   More plain text

<RXANDCLEAR>   This is a command to return the radio to the receive mode and clear (erase) the transmitted text from the transmit buffer so that it won't accidentally be sent again.

The T/R button executes a single macro command.

As you can probably see, the T/R button is also a macro. When you click on it, the **TXTOGGLE** command executes. This places the radio into transmit mode and then waits for you to click the button again to return to the receive mode.

If you right click on one of the macro buttons again and browse the scrolling list of commands along the right-hand side of the window, you'll quickly realize that you can create some incredibly elaborate macros. Contest macros, for example, can automatically send the other station's call sign, a signal report and a serial number that increments by one with every contact – all with a single click of your mouse.

Be careful about relying too much on macros. They are convenient – especially the CQ macro – but they can also cause confusion if used improperly. In addition, some hams create macros that contain lengthy descriptions of their station equipment, families, etc. It's tempting to tap that macro key and sit back while your computer sends more information than the other person really wants to know!

Finally, many amateurs complain – and with some justification – that overreliance on macros takes the pleasure out of making contacts. A contact based on macros is little more than a brief exchange of canned text. A ham who is expecting a leisurely chat will be gravely disappointed if all you do is send a few macros instead.

Chapter Three

RTTY

The Short Scoop
Radioteletype is an old digital mode, but
still very popular today, especially for
contesting and DXing.

Radioteletype, better known as *RTTY*, is the "old man" of the HF digital modes. Despite more than six decades of amateur use it is still the top mode for HF digital contesting. It also remains the king of HF digital DXpeditions. There are several reasons for this …

•*Sheer momentum.* RTTY has a rich, long history and a large number of HF digital operators have grown comfortable with it over the years. Like CW, RTTY has a venerated place in Amateur Radio lore.

•*RTTY-friendly transceivers.* HF transceivers have been designed with RTTY in mind for decades. As you may recall from Chapter 1, a substantial number of RTTY operators still prefer to use FSK and transceiver manufacturers continue to make this mode available in their designs. (If you don't know the meaning of FSK, and why it is particularly relevant to RTTY, review Chapter 1.)

•*Performance.* Several HF digital modes are superior to RTTY when it comes to spectrum efficiency and weak-signal performance, but RTTY still has an ace up its sleeve. In crowded conditions such as contests and DX pileups, RTTY can often be decoded when other modes cannot. PSK31, for example, has a "capture effect" where only the stronger of two signals on the same frequency can be decoded. In contrast, RTTY is capable of coming up with a partial print of two signals as they battle it out on the same frequency, depending on their relative strengths. This is important when you're a contest or DX station and several RTTY operators are calling you at once.

•*Speed.* Coupled with its ability to tolerate crowded conditions is RTTY's 60 word-per-minute speed. That's about as fast as many of us can type, so it is more than adequate. Other digital modes offer better weak-signal performance, but they often include a speed penalty. Not so for RTTY.

RTTY is very easy to use and every multimode digital software package listed in Chapter 1 includes RTTY in its collection. Also, there is a free, high-performance RTTY program known as *MMTTY* by Makoto Mori, JE3HHT. In this chapter I'll guide you through a RTTY contact using *MMTTY*.

Before we get started, however, let's spend some time becoming familiar with a few important RTTY terms.

Mark, Space, Shift and Baud

RTTY uses an old digital code known as *Baudot*. Each Baudot character is composed of 5 bits. That's enough to send every letter in the English alphabet, plus the number 0 through 9 and some limited punctuation.

In amateur RTTY communication a "1" bit is represented by a 2125-Hz tone and this is known as a *mark* signal. A "0" bit is represented by a 2295-Hz tone called a *space* signal. There is also a start pulse at the beginning of the bit string and a stop pulse at the end (see **Figure 3.1**). The rapid switching between mark and space tones gives RTTY its distinctive *deedle-deedle* sound. When you see a RTTY signal in a waterfall display, you can instantly spot the twin parallel lines representing the mark and space signals. See **Figure 3.2**.

As you examine Figure 3.2, you'll notice that the signal lines are separated by an "empty" area. This separation isn't random. In fact, it is highly precise. Grab your calculator and do a bit of subtraction:

$$2295 \text{ Hz} - 2125 \text{ Hz} = 170 \text{ Hz}$$

In the answer shown above, 170 Hz is the difference or *shift* between the mark and space frequencies. The Amateur Radio RTTY standard is to use a 170-Hz shift. Maintaining proper shift is critical to RTTY communication.

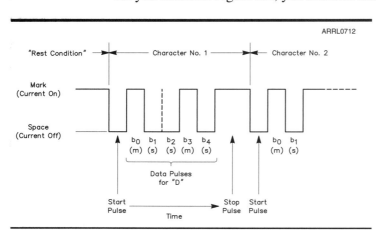

Figure 3.1 – This is a diagram of a RTTY signal as the letter "D" is sent in Baudot code. A start pulse begins the character, followed by the five bits (b0 – b4) that define it. A stop pulse signals the end of the character.

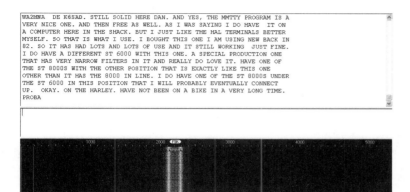

WA2MWA DE K6SAD. STILL SOLID HERE DAN. AND YES, THE MMTTY PROGRAM IS A
VERY NICE ONE. AND THEN FREE AS WELL. AS I WAS SAYING I DO HAVE IT ON
A COMPUTER HERE IN THE SHACK. BUT I JUST LIKE THE HAL TERMINALS BETTER
MYSELF. SO THAT IS WHAT I USE. I BOUGHT THIS ONE I AM USING NEW BACK IN
82. SO IT HAS HAD LOTS AND LOTS OF USE AND IT STILL WORKING JUST FINE.
I DO HAVE A DIFFERENT ST 6000 WITH THIS ONE. A SPECIAL PRODUCTION ONE
THAT HAS VERY NARROW FILTERS IN IT AND REALLY DO LOVE IT. HAVE ONE OF
THE ST 8000S WITH THE OTHER POSITION THAT IS EXACTLY LIKE THIS ONE
OTHER THAN IT HAS THE 8000 IN LINE. I DO HAVE ONE OF THE ST 8000S UNDER
THE ST 6000 IN THIS POSITION THAT I WILL PROBABLY EVENTUALLY CONNECT
UP. OKAY. ON THE HARLEY. HAVE NOT BEEN ON A BIKE IN A VERY LONG TIME.
PROBA

Figure 3.2 – This is *MixW* multimode software receiving a RTTY signal. You can clearly see the twin lines that represent the *mark* and *space* tones. The separation between mark and space is known as the *shift*.

RTTY software that's configured for a 170 Hz shift creates narrow DSP filters that "expect" this exact frequency separation. A RTTY signal that's using a different shift will fall outside the filters and will be undecodable. FCC rules limit amateur RTTY to a 1 kHz maximum shift, but you'll rarely encounter anything other than a 170 Hz shift.

You'll also hear occasional references to a RTTY signal being "inverted." When this occurs, the mark and space tone frequencies are swapped. As long as the station you are communicating with is also inverted, this isn't a problem. However, it will render your text as gibberish to all the other RTTY operators who are "right side up."

When it comes to the signaling rate (or "speed," if you will), the vast majority of RTTY you'll hear on the air will be perking along at a rate of 45 baud (60 words per minute). That said, you will occasionally run into 75 baud RTTY.

On the Air with RTTY – Step by Step

If you don't already have a multimode program that offers RTTY, give *MMTTY* a try. You can download it free at **http://hamsoft.ca/pages/mmtty.php**, or grab it at the special *Get on the Air with HF Digital* page on the ARRL website at **www.arrl.org/arrl-store/HF-Digital**. You will also need to download *EXTFSK*, a small COM port driver program for use with *MMTTY*, which you'll find at **http://hamsoft.ca/pages/mmtty/ext-fsk.php** (scroll down to the bottom of the page), or at the *Get on the Air with HF Digital* page.

Important note: Install *MMTTY* first. After installing *MMTTY*, extract the *EXTFSK* files ***to the directory where MMTTY resides***.

As we did with PSK31 in Chapter 2, let's use *MMTTY* as an example and step through setting up the software. *MMTTY* is somewhat complex because it offers the ability to tweak many of its performance parameters. Even so, its *default* settings – the ones already in effect when you install the software – are

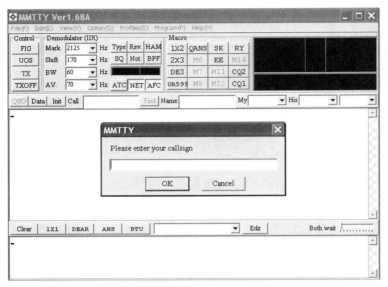

When you run *MMTTY* for the first time, you'll be asked to enter your call sign.

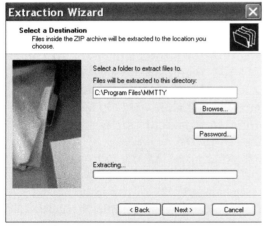

Install *MMTTY* first, then download *EXTFSK*. Make sure to extract the files into the *MMTTY* directory on your hard drive as shown here.

perfectly adequate for our purposes. If you are using different RTTY software, you'll find that the *MMTTY* RTTY configuration is quite similar.

MMTTY Setup

Start by clicking your mouse cursor on OPTION(O) along the top of the *MMTTY* window. Now click on SETUP MMTTY(O).

A Setup window will open with seven tabs arrayed horizontally. All these tweakable settings may look intimidating, but we're only concerned with a few.

Click on the SOUNDCARD tab at the far right. You'll be presented with a list of the sound devices that *Windows* has detected. Just click on the empty circles next to the sound device you are using for receive audio from your radio (audio input to the sound device) and transmit audio (audio output) going to your radio. Do *not* click OK (at least not yet).

Click on the TX tab. In the upper right corner of this window you'll see an area labeled PTT & FSK. Click on the arrow in the PORT window and highlight EXTFSK at the very bottom of the drop-down menu. If *EXTFSK* is not showing up in the PORT menu that means *EXTFSK* was not properly placed in the same directory on your computer as *MMTTY.EXE*.

After *EXTFSK* has been selected, go to the MISC tab and select either SOUND

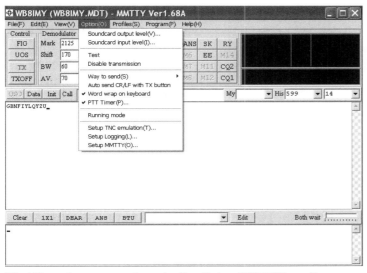

Start the setup process by selecting Setup *MMTTY* from the Option(O) menu.

The *MMTTY* setup window. Notice the tabs along the top.

+ COM-TXD (FSK) if you are using your transceiver microphone or accessory jack as the transmit audio input, or COM-TXD (FSK) if you've set up your radio for "true" FSK keying as described in Chapter 1. If you are just dabbling in RTTY for the first time and you've already set up your station for PSK31 or one of the other sound-device-based modes, select SOUND + COM-TXD (FSK). Now click on the USB PORT button to open the USB PORT OPTION window. Select NORMAL in this window and click OK.

Go back to the TX tab and click the OK button. The *EXTFSK* window will appear separately from the *MMTTY* window. In this window, you can select which COM port you wish to configure. Once again, if you are using your transceiver microphone or accessory jack as the transmit audio input, select either the RTS or DTR line so that *EXTFSK* can key your transceiver's PTT (Push to Talk) line at the microphone or accessory jack.

In most cases the correct choice will probably

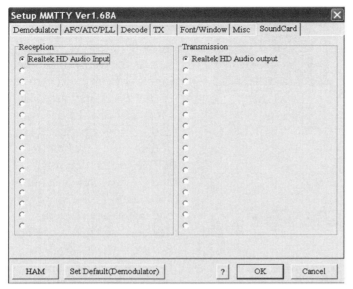

Select your sound device in this *MMTTY* window.

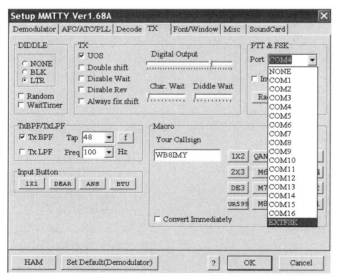

In this menu you'll select EXTFSK as your driver for transmit/receive switching.

be DTR. On the other hand, if you are using FSK for RTTY, you may need to select TxD. If you're in doubt, consult your interface manual. At worst, you may need to experiment until you find which keying line does the trick. The good news about setting up *EXTFSK* is that you only have to do it once.

The *EXTFSK* window will remain open while you're using *MMTTY*. If necessary, just drag it off to the side and out of the way. When you close *MMTTY*, it will close as well; it will re-appear when you open *MMTTY* again.

Set Up Your Transceiver

Now that you've set up the important functions of *MMTTY*, it's time to bring your radio into the picture. What happens next depends on your transceiver design.

As we discussed in Chapter 1, if your transceiver has a "RTTY," "Digital" or "Data" mode this can mean different things to different manufacturers. It could be little more than a means to select the incoming and outgoing audio pathways and possibly implement some useful filter options. On the other hand, it could be true FSK where the radio is expecting data pulses (not audio) from your computer

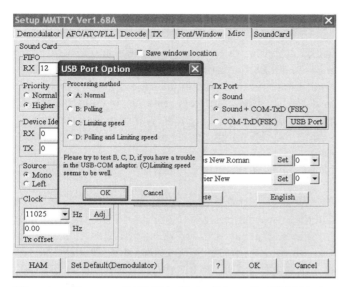

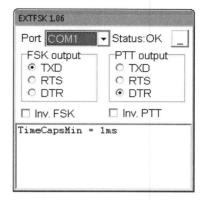

The *EXTFSK* setup window. This is where you finally choose the COM port that will switch your rig between transmit and receive.

After choosing your TX PORT option, click on the USB PORT button and select NORMAL as your processing option, then click OK.

or interface. You'll need to consult your transceiver manual to be sure. For the sake of moving forward, let's assume that your DIGITAL mode involves taking audio to and from your interface and software. Turn on the transceiver and select your "Digital" mode now.

If your radio doesn't have a "Digital" mode, select LOWER SIDEBAND (LSB) instead. This is the standard for AFSK RTTY when using SSB transceivers. Selecting USB will flip your mark/space tone relationships upside down. As a result, you won't be able to decode signals, and other stations won't be able to decode your transmissions.

Take a moment to test the transmitter. Tune your radio to the middle of one of the frequency ranges in **Table 3.1** and make sure the frequency is clear. If not, pick another. Once you are in a clear spot, click your mouse cursor on the red-labeled *MMTTY* TX button. Your transceiver should switch to the transmit mode.

If you are using AFSK RTTY, you need to be careful about overdriving your radio with transmit audio. As with PSK31, this will generate a terribly distorted signal. While transmitting, switch your transceiver meter to ALC. If the meter is showing ALC activity, or if the ALC indicator is swinging out of the "safe" zone as described in Chapter 1, reduce your transmit audio. Depending on the type of meter your radio offers, there should be no ALC activity, or the needle should never leave the safe zone.

Table 3.1
Common RTTY Frequencies
3580 – 3600 kHz
7040 – 7100 kHz
14080 – 14099 kHz
21080 – 21100 kHz
28080 – 28100 kHz

If you are using true FSK, you're spared from worrying about any of this. That's one of the convenient features of FSK! Your interface is sending data pulses to your transceiver and your transceiver automatically generates the proper distortion-free RF output.

This is a good time to remind you about RF output power. Remember that RTTY is a 100% duty cycle mode. If your rig is designed to handle 100% duty cycle signals, you're safe. Otherwise, you'll need to reduce your RF output by as much as 50%. Check your transceiver manual.

Click the TXOFF button to return to receive.

Tuning and Decoding RTTY Signals

Every RTTY decoder, whether it works in software or hardware, incorporates a set of mark and space audio filters. You can think of these filters as dual windows that only open for tones that are at the correct mark and space frequencies, and separated by the proper shift. The mark and space filtering circuitry detects and decodes the tones into digital 1s and 0s, which is exactly what your computer needs to provide text on your screen. The more sensitive and selective your mark/space detectors, the better your RTTY performance, especially as it involves your ability to copy weak signals through interference.

It's easy to understand why tuning a RTTY signal (and just about any other data signal) is so critical—and why a good tuning indicator is one of your best HF digital tools. Whenever you stumble upon a RTTY signal, you must quickly tune your receiver until its mark and space tones fall within the "skirts" of the filters and are detected. It's possible to do this by ear once you become accustomed to the sound of a properly tuned RTTY transmission, but few of us have the necessary patience. Instead, we rely on visual indicators to guide us.

In *MMTTY* you have two default indicators in

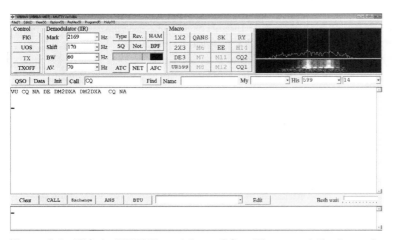

Figure 3.3 – This is *MMTTY* receiving a CQ call from a relatively weak station. Notice how the waveform peaks on the two mark and space tuning lines. The mark and space signals also appear as bright lines in the waterfall display below.

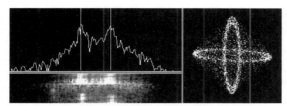

Figure 3.4 – The *MMTTY* tuning indicator with the XY Scope display activated. In addition to the sharp mark and space peaks in the spectral display, the ellipses are crossed in the scope window.

the upper right corner of the screen. The top section displays the receive audio as a rapidly varying waveform. Just below this window is a grayscale waterfall display. As you can see in **Figure 3.3**, a RTTY signal appears as two peaked waveforms when tuned properly in the upper window. These are the mark and space signals. Notice how the peaks are centered on the two white tuning lines. In the waterfall display the mark and space signals appear as two white traces directly beneath the tuning lines.

One of the strengths of *MMTTY* is its flexibility and that applies to the tuning indicator as well. In the **View (V)** menu you can choose to activate the **XY Scope** (see **Figure 3.4**). This is a software variation of what RTTY old timers used to call the "crossed bananas" display. You tune the RTTY signal until the ellipses cross.

Unless there is a contest or a DX pileup going on, finding RTTY on the air may require some patience. Take another look at the frequencies in Table 3.1. This is where you are most likely to encounter RTTY on any given day. Twenty meters is the most popular band. Another option is to monitor a W1AW RTTY transmission from ARRL Headquarters. You'll find the W1AW bulletin schedule online at **www.arrl.org/digital-transmissions**. Note that RTTY transmissions are referred to as "Baudot" in the W1AW schedule.

Once you've found a signal, tune it in carefully and watch the text march across the large receive window. *MMTTY* is sensitive software and it has a tendency to respond to random noise by printing equally random characters in the receive window. If this annoys you, active the software squelch function by clicking your mouse cursor on the SQ button.

As you watch the signal, notice how the conversation flows just like a voice or CW ragchew.

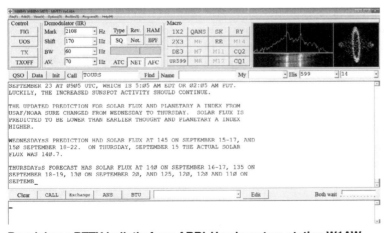

Receiving a RTTY bulletin from ARRL Headquarters station W1AW.

KF6I DE WB8IMY . . . YES, I HEARD FROM SAM JUST YESTERDAY. HE SAID THAT HIS TOWER PROJECT WAS ALMOST FINISHED. KF6I DE WB8IMY K

If you want to call CQ, the procedure is simple. *MMTTY* has two "canned" CQ messages (known as *macros*) that you can send by clicking on either the CQ1 or CQ2 buttons. You can right click on these buttons to change the macros to your liking. See the discussion of macros in Chapter 2.

Alternatively, you can manually type your CQ in the bottom transmit buffer window. When you click the TX button, *MMTTY* will send everything in the transmit buffer.

For now, let's try the CQ1 button. Click on it and you'll hear the delightful chatter of RTTY and see the following on your screen (with your call sign instead of mine):

CQ CQ CQ DE WB8IMY WB8IMY WB8IMY PSE K

Notice how the CQ is short and to the point. Did you also notice that it repeated the call sign several times? *Remember that RTTY lacks error detection.* If you want to make certain that the other station copied what you sent, it helps to repeat it. You'll see this often in contest exchanges. For example:

N6ATQ DE N1RL. . . UR 549 549. STATE IS CT CT. DE N1RL K

On the other hand, if you know that the other station is copying you well, there is no need to repeat information.

An option to investigate in your RTTY software is Unshift On Space (UOS). There is a benefit to using UOS when receiving. Two special characters, LTRS and FIGS, are used by the RTTY code to indicate to the computer or processor whether the characters that follow will letters or figures (numbers and punctuation). If you enable UOS, your software will exit the FIGS shift as soon as there is a space, so your display does not print out garbage if the software misses the LTRS shift character because of noise or interference. In *MMTTY* USO is the default.

MMTTY has an Automatic Frequency Control (AFC) function that will attempt to track the received signal as it drifts due to propagation conditions or other factors. You can disable this by clicking on the AFC button. The adjacent NET button has a similar effect on your transmitted signal.

AFC and NET are both handy functions, but they can cause problems when not used correctly. If you are enjoying a casual RTTY conversation on an uncrowded band, you can leave AFC and NET on. But if the band is crowded – during a RTTY contest, for example – you may be better off turning both functions off. Many RTTY operators run very tight filters in crowded band conditions, some as narrow as 250 Hz. If the AFC and NET are on, *MMTTY* could easily shift your signal outside the range of the other station's filters. By the same token, you could be calling CQ with a narrow IF filter and wondering why no one replies!

Chasing RTTY DX

RTTY is still the most widely used mode for HF digital DXing. You can work 100 countries to earn your RTTY DX Century Club (DXCC) award, and even climb the ladder to the stratospheric heights of RTTY DXCC Honor Roll. RTTY DXing is competitive and highly rewarding.

Like any other form of DXing, the quest for RTTY DX demands patience and skill. When a DXpedition is on the air with RTTY from a rare DXCC entity, your signal will be in competition with thousands of other HF digital operators who want to work the station as badly as you do. Sometimes pure luck is the winning factor, but there are several tricks of the trade that you can use to tweak the odds in your favor.

Don't Call...Yet

Let's say that you're tuning through the HF digital subbands one day and you stumble across a screaming mass of RTTY signals. On your computer screen you see that everyone seems to be frantically calling a DX station. This is known as a *pileup*.

You can't actually hear the DX station that has everyone so excited, but what the heck, you'll activate your transceiver and throw your call sign into the fray, right? *Wrong!*

Never transmit even a microwatt of RF until you can copy the DX station. Tossing your call sign in blindly is pointless and will only add to the pandemonium. Instead, take a deep breath and wait. When the calls subside, can you see text from the DX station on your screen? If not, the station is probably too weak for you to work (don't even bother), or he may be working "split." More about that in a moment.

If you can copy the DX station, watch the exchange carefully. Is he calling for certain stations only? In other words, is he sending instructions such as "North America only"? Calling in direct violation of the DX station's instructions is a good way to get yourself black-listed in his log. (No QSL card for you—ever!) Does he just want signal reports, or is he in the mood for brief chats? Most DX stations simply want "599" and possibly your location—period. Don't give them more than they are asking for. (A DX RTTY station on a rare island doesn't care what kind weather you are experiencing at the moment.)

Working the Split

When DX RTTY pileups threaten to spin out of control, many DX operators will resort to working *split*. In this case, "split" means split frequency. The DX station will transmit on one frequency while listening for calls on another frequency (or range of frequencies).

A good DX operator will announce the fact that he is working split with almost every exchange. That's why it is so important to listen to a pileup before you throw yourself into the middle. If you tune into a pileup and cannot hear the DX station, tune below the pileup and see if you copy the DX station there. If his signal is strong enough, he shouldn't be hard to find if he is working split. His signal will seem to be by itself, answering calls that you cannot hear. This is a major clue that a split operation is taking place. Watch for copy such as…

CQ DX DE FOØAAA, UP 10 (Translation: He is listening up 10 kHz)

CQ DX DE FOØAAA, 14085-14090 (Translation: He is listening between 14085 and 14090 kHz)

Whatever you do, *never call a split DX operation on the station's transmitting frequency*. Your screen will quickly fill with rude comments from others who are listening. Instead, put your transceiver into the split-frequency mode (better drag out your user manual if you're unsure how to do this). Set your radio to receive on the DX station's transmitting frequency and transmit on his listening frequency. If the DX station is listening through a range of frequencies, you'll need to select a spot where you think you'll be heard. Change your transmit frequency if this particular "fishing spot" doesn't seem to be working.

Short and Sweet

When you've done your listening homework and you're ready for battle, by all means fire at will. Wait until the DX station finishes an exchange. He'll signal that he is ready for another call by sending "QRZ?" or something similar. When it's time to transmit, make it short and to the point, like this…

WB8IMY WB8IMY WB8IMY K

Listen again. Has he responded to anyone yet? If the answer is "no" and other stations are still calling, the DX is probably trying to sort out the alphabet soup of confusion on his screen. Try again.

WB8IMY WB8IMY WB8IMY KK

You might make it through at just the right moment when other signals subside briefly, or when the ionosphere gives you an unexpected boost. But if no one is calling, or if you hear the DX station calling someone else, *stop*. You lost this round, so give the lucky winner his chance to be heard. Your next opportunity will be coming up shortly.

Did He Call You?

Watch your screen carefully. If the DX station only copied a fragment of your call sign, he might send something like…

IMY IMY AGAIN??

In this instance he copied only the last three letters of my call sign ("IMY"). Fading and flutter can make RTTY signals difficult to copy clearly. The best thing to do is reply right away, sending your call sign three times just like before.

If luck is on your side, you'll see…

WB8IMY DE FOØAAA . . . TNX. 599. QSL? K

And with excited fingers you answer…

QSL. UR 599. TNX AND 73 DE WB8IMY K

RTTY Contesting

Nothing tests your equipment and operating skills like a contest, and when it comes to digital contesting RTTY is the #1 mode. See the list of major annual RTTY contests in **Table 3.2**.

A contest really shakes the bugs out of your station. If you have shortcomings in your antenna system, you're going to discover them very quickly. If your transceiver can't seem to handle the crunch of dozens of signals in proximity, it will become painfully apparent within minutes. If you need better logging or contest

Table 3.2
Major RTTY Contests Throughout the Year
For details, see the ARRL website at **www.arrl.org/contest-calendar** *or the SM3CER contest page at* **www.sk3bg.se/contest/.**

New Year's Day	SARTG New Year RTTY Contest
First weekend in January	ARRL RTTY Roundup
Third weekend in January	BARTG RTTY Sprint
Second weekend in February	CQ World Wide WPX RTTY Contest
Fourth weekend in February	North American QSO Party
Second weekend in March	North American RTTY Sprint
Third weekend in March	BARTG HF RTTY Contest
First weekend in April	EA RTTY Contest
Second weekend in May	A.Volta RTTY DX Contest
Third weekend in July	North America QSO Party
Third weekend in August	SARTG RTTY Contest
Last weekend in September	CQ WW RTTY DX Contest
Second weekend in October	BARTG RTTY SPRINT
Third weekend in October	JARTS World Wide RTTY Contest
Second weekend in November	Worked All Europe DX Contest
Third weekend in December	OK DX RTTY Contest

software, the first hour of a contest will provide a powerful motivation to upgrade.

Some hams shun contesting because they assume they don't have the time or hardware necessary to win—and they are probably right. But winning is *not* the objective for most contesters. You enter a contest to do the best you can, to push yourself and your station to whatever limits you wish. The satisfaction at the end of a contest comes from the knowledge that you were part of the glorious frenzy, and that you gave it your best shot.

Contesting also has a practical benefit. If you're an award chaser, you can work many desirable stations during an active contest. During the ARRL RTTY Roundup, for example, some hams have worked enough international stations to earn a RTTY DXCC. Jump into the North American RTTY QSO Party and you stand a good chance of earning your Worked All States in a single weekend.

Always remember that contesting is ultimately about *enjoyment*. The thrill of the contest chase gets your heart pumping. The little triumphs, like working that distant station even though he was deep in the noise, will bring smiles to your face.

You'll make a number of friends along the way, too. Why do you think it is called the "brotherhood of contesting"?

Digital Contesting Tips for the "Little Pistol"

If you are like most amateurs, your station falls into the Little Pistol category. Mine does, too. The Little Pistol station usually includes a low- to medium-priced 100-W transceiver feeding an omnidirectional antenna such as a dipole, vertical, end-fed wire and so on. Even if you are blessed with a beam antenna on a tower or rooftop, in all likelihood you are still a Little Pistol.

At the opposite end of the spectrum are the Big Gun stations. They have multiple beams on tall towers, phased array antennas on the lower bands, incredibly long Beverage receiving antennas and more. They own high-end (read: expensive) transceivers that boast razor-sharp selectivity, and they usually couple them to 1.5 kW linear amplifiers.

So how does a Little Pistol go about beating a Big Gun in a digital contest? With rare exceptions, he can't. Banging heads with a Big Gun station is usually an exercise in futility. He is going to hear more stations, and more stations are going to hear him. Focus, instead, on using your wits and equipment to get the best score possible. You probably won't beat a Big Gun, but you could make him sweat!

Check Your Antenna

How long has it been since you've inspected your antenna? Is everything clean and tight? And have you considered trying a different antenna? Contests are terrific testing environments for antenna designs. For example, if you've been using a dipole for a while, why not try a loop? Just set up the antenna temporarily for the duration of the contest and see what happens.

Check Your Equipment

If you are having problems with RF getting into your computer or sound card, fix these glitches *before* you find yourself in a contest. Also, consider spending $140 or so to install a 500-Hz IF filter in your radio if it doesn't have one already, or if it lacks built-in IF DSP filtering. Finally, check your computer and software. Make sure you understand the program completely. Set up any "canned" messages/responses and test them before the contest begins.

Understand Propagation

If you are participating in a DX contest, is it a good idea to prowl the 40-meter band in the middle of the day? No, it isn't. Forty meters is only good for DX contacts after sundown. If you are hunting Europeans or Africans on 10 or 15 meters, when is the best time to look for them? The answer is late morning to early afternoon.

In other words, let propagation conditions guide your contest strategy. Your goal is to squeeze the most out of every band at the right time of day. For example, I might start the ARRL RTTY Roundup on 10 meters and then bounce between 10, 15 and 20 meters until sunset. After sundown, I'll concentrate on 20 and 40 meters. In the late night and the wee hours of the morning, I'll probably limit myself to 40 and 80 meters.

In some contests you can enter as a single-band station. Which band should you choose? That depends on the contest and your equipment. If you have a 15-meter beam and you are entering a DX RTTY contest, it makes sense to concentrate on 15 meters, even though you'll probably find yourself with nothing to do in the late evening after the band shuts down. Twenty meters makes sense as the ultimate "around the clock" band, but it is often very crowded during contests and populated with Big Gun stations. In some instances you may find it more "profitable" to concentrate your efforts on another band without so many signals.

Hunting and Pouncing vs. "Running"

The common sense rule of thumb is that a Little Pistol station should only hunt and pounce. That means that you patrol the bands, watching for "CQ CONTEST" on your monitor, and pouncing on any signals you find. The Big Guns, on the other hand, often set up shop on clear frequencies and start blasting CQs. If conditions are favorable, they'll be hauling in contacts like a commercial fishing trawler! This is known as *running*.

In many cases Little Pistols are probably wasting valuable time by attempting to run. There are situations, however, where running *does* make sense. If you've pounced on every signal you can find on a particular band, try sending a number of CQs yourself to catch some of the other pouncers. If you send five CQs in a row and no one responds, don't bother to continue. Move to another band and resume pouncing.

Seek the Multipliers

Every contest has multipliers. These are US states, DXCC countries, ARRL sections, grid squares and so on, depending on the rules of the contest. A multiplier is valuable because it multiplies your total score.

Let's say that DXCC countries are multipliers for our hypothetical contest. You've amassed a total of 200 points so far, and in doing so you made contacts with 50 different DXCC countries.

$$200 \times 50 = 10,000 \text{ points}$$

Those 50 multipliers made a huge difference in your score! Imagine what the score would have been if you had only worked 10 multipliers?

If given a choice between chasing a station that won't give me a new multiplier and pursuing one that *will*, I'll spend much more time trying to bag the new multiplier.

Contest Software

No one says you have to use software to keep track of your contest contacts, but it certainly makes life easier! One of the fundamental elements of any contest program is the ability to check for duplicate contacts or *dupes*. Working the same station that you just worked an hour ago is not only embarrassing, it is a waste of time. The better contest programs feature automatic dupe checking. When you

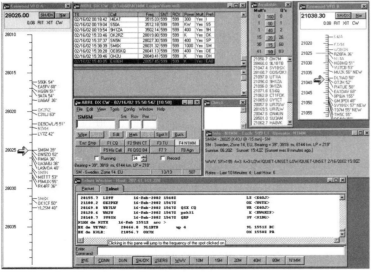

N1MM contest logging software.

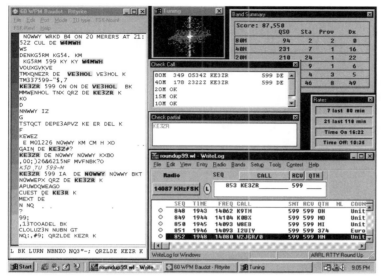

WriteLog in action.

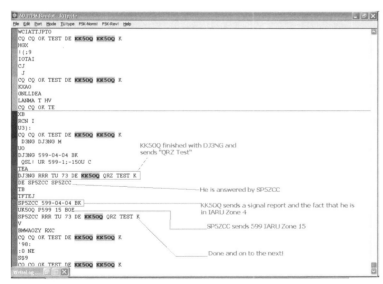

A typical contact contact as shown in **WriteLog** with labels added for clarity. Here KK5OQ is "running" and working stations one after another.

enter the call sign in the logging window, the software instantly checks your log and warns you if the contact qualifies as a dupe. The more sophisticated programs "know" the rules of all the popular digital contests and they can quickly determine whether a contact is truly a dupe under the rules of the contest in question. Some contests, for example, allow you to work stations only *once*, regardless of the band. Other contests will allow you to work stations once per band.

A good software package will also help you track multipliers. It will display a list of multipliers you've worked, or show the ones you still need to find.

Any of the contest software packages advertised in *QST* and the *National Contest Journal* (*NCJ*) will function well. To minimize headaches, however, my suggestion is to stick with software written specifically for contesting. There are two popular contesting software packages on the market today:

•*N1MM Logger* is a popular free contesting application for *Windows*. You can download the latest version at **http://n1mm.hamdocs.com/**. *N1MM Logger* supports RTTY contesting with *MMTTY*.

•*WriteLog* by Ron Stailey, K5DJ, is a full-featured *Windows*-based software package with an interesting twist – a built-in RTTY program. Like *RTTY*, *WriteLog* offers automatic dupe checking and flagging, multiplier displays, radio control, canned messages and more. See the *WriteLog* site on the Web at **www. writelog.com**.

Chapter Four

JT65

The Short Scoop
JT65 isn't a mode for conversation, but when it comes to making valid contacts with low power and poor antennas, JT65 is hard to beat.

"I'm hearing a strange signal on several HF bands. It sounds like someone sending random tones or music. It plays slowly for about 50 seconds, stops for a while, and then plays again. What is it?"

At the time this book was written, this question was arriving in e-mail inboxes at the ARRL Headquarters Regulatory Branch at a rate of about once per week. Some amateurs even send audio recordings of the suspicious signals. Chuck Skolaut, KØBOG, our Field and Regulatory Correspondent, always smiles as he listens to the recordings because he knows the answer by heart: *JT65*.

By now amateurs are used to the sounds most digital modes create. They've learned to recognize the constant warbling tones of PSK31, the rhythmic pulses of PACTOR, the "scratchy" rumble of Hellschreiber or the multi-tone music of RTTY, MFSK16, Olivia and others.

But JT65 is unique. It marches, as Thoreau said, to the beat of a different drummer. If you've never heard it before it will stop you cold. As you tune across a JT65 signal you'll hear tones of varying pitch that "play" slowly, like someone lazily pecking on an electronic keyboard.

Cryptic and strange as the tones may be, you might be surprised to learn that they carry call signs, signal reports and other bits of information. Even more surprising is the fact that information can be extracted from a JT65 signal even when it is extremely weak.

The "JT" in JT65

JT65 debuted as part of the *WSJT* software suite created by Dr Joe Taylor, K1JT. As a Nobel Prize-winning scientist who studies pulsars and other distant astronomical objects, Joe has a keen interest in weak signals. Joe's software exploits the power of modern desktop and laptop computers to separate weak signals from noise and decode the information they contain. With just a sound card or sound chipset and a transceiver interface, *WSJT* makes it possible for hams with modest stations to enjoy VHF meteor scatter communication and even moonbounce, where signals are literally bounced off the surface of the Moon and returned to Earth. *WSJT* is available for free downloading at **http://physics. princeton.edu/pulsar/K1JT/Download.htm**.

In the beginning, JT65 was embraced by some members of the moonbounce community and it was an instant success. Thanks to JT65, amateurs with single long-boom Yagi antennas and 150 W of RF output can experience the thrill of communicating over the longest "Long Path" of all.

But it wasn't long before someone wondered what would happen if JT65 was used on the HF bands. Digital communication on HF isn't nearly as challenging as getting a signal to the Moon and back, so it stood to reason that there would be plenty of "performance margin" to provide fascinating results. To no one's surprise, this turned out to be true. Using a variant of JT65 known as JT65A, even a few watts of JT65 modulated RF to a wire dipole antenna resulted in transcontinental and even global communication.

Dedicated JT65 Software

JT65 is one of several modes in the *WSJT* package, available for free downloading on the web at **http://physics. princeton.edu/pulsar/ K1JT/Download.htm**. However, Joe Large, W6CQZ, thought more amateurs might try JT65 on HF if it was available in software specifically designed to make it easier

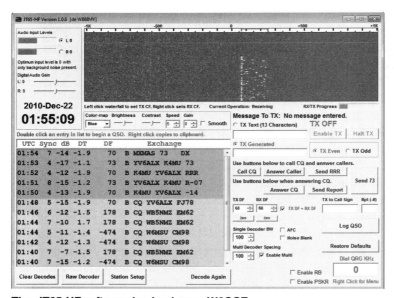

The *JT65-HF* software by Joe Large, W6CQZ.

to enjoy contacts. The result was his *JT65-HF* application for *Windows* and it soon proved Joe's hunch correct. Since Joe's software made its debut, JT65 activity on the HF bands has increased substantially. More about Joe's amazing software in a moment.

So What is JT65 Anyway?

The short and simplified answer to this question is that JT65 is a weak-signal digital mode that uses one-minute transmit/receive sequences, meaning that you transmit within a one-minute window and then listen for one minute. Transmission actually begins one second after the start of a UTC minute and stops precisely 47.7 seconds later. There is a 1270.5 Hz synchronizing tone and 64 other tones. This combination gives JT65 its unusual musical quality.

Tight time and frequency synchronization is critical to JT65. Your SSB transceiver needs to be reasonably stable, although I've yet to see a modern commercial radio that is too "drifty" for JT65. Drifty computer time is a different matter, however. *Windows* PCs are notorious for sloppy timekeeping, but there are ways to deal with this, as you'll see.

JT65 is not a "conversational" mode like, say, PSK31. Instead, the idea is to exchange just the basic information required for a valid contact: call signs and signal reports. *JT65-HF* measures the actual received signal strength and incorporates it into the exchange. When you receive a report during a *JT65-HF* exchange, you'll know exactly how strong your signal is (in dB) at the other end.

JT65 contacts count for many awards such as Worked All States or the DX Century Club. Aside from the fun of award chasing, it is amazing to see who you can contact with JT65 while using miniscule amounts of power. Some JT65 enthusiasts are using output levels in the *milliwatt* range. In fact, 50 W is considered "high power" in the JT65 world.

JT65-HF Software Setup – Step by Step

JT65 is an AFSK mode in which you apply transmit audio to your transceiver, just like PSK31 and other sound-device-based modes. If you've set up your station accordingly, as described in Chapter 1, you're good to go.

Most amateurs trying JT65 on the HF bands are using W6CQZ's software and I recommend that you do the same. You can download the latest version at no charge from Joe's SourceForge page at **http://sourceforge.net/projects/JT65-HF/files/**. Or you can get it from the special *Get On the Air with HF Digital* page on the ARRL Web at **www.arrl.org/HF-Digital**. There isn't a version of *JT65-HF* for Mac or *Linux* users, but I have managed to run *JT65-HF* successfully under *Linux* using *Wine*. Mac and *Linux* users may also be able to do the same by

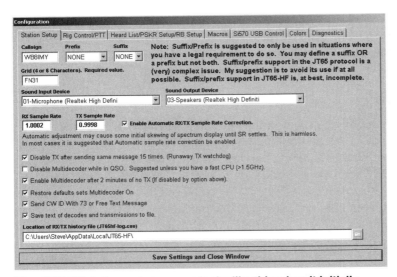

The *JT65-HF* station setup screen looks like this when it initially opens. Just add your call sign and grid square, and make sure the correct sound devices are selected.

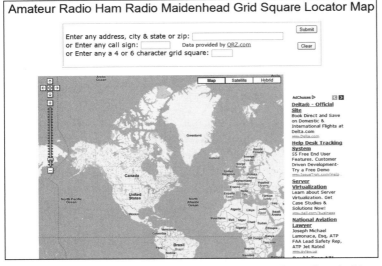

K2DSL's grid square calculator at www.levinecentral.com/ham/grid_square.php.

using a *Windows* emulator such as *CrossOver* (**www.codeweavers.com**).

Before you start *JT65-HF*, make sure your interface is plugged into your computer. *JT65-HF* is one of those applications that needs to "see" that the audio inputs and outputs are properly in place when it starts or it will generate a cryptic error message.

When you start *JT65-HF* for the first time, your first task is to set up the program. Click on SETUP in the upper left corner. This will open the CONFIGURATION window. This window has seven tabs lined up horizontally, but we only need to be concerned about a few of them.

The most important tab is STATION SETUP and it should be visible to you immediately. Fill in your call sign and your grid square. If you don't know your grid square, you can find out online by using K2DSL's handy calculator at **www.levinecentral.com/ham/grid_square.php**. All you have to do is enter your ZIP code and the site will respond with your grid square.

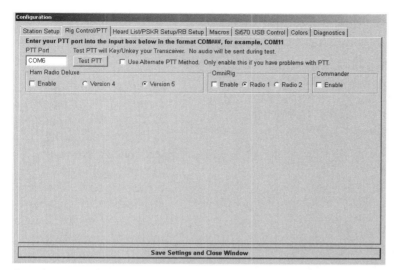

On this portion of the STATION SETUP screen you enter the COM port for your computer/transceiver interface.

Select your SOUND INPUT and SOUND OUTPUT devices. Click the drop-down arrows and you'll see a list of every device *Windows* recognizes. If you are using the sound card or sound chipset in your computer it will appear in this list, probably with a label such as "Microphone" or "Speakers." If you are using an interface with a built-in sound device, it may appear with an odd label such as "USB Codec."

I suggest that you accept the default settings for all the checkboxes that appear below this section, particularly the one labeled ENABLE AUTOMATIC RX/TX SAMPLE RATE CORRECTION. This handy feature will automatically adjust your sound device sample rate when the program starts. You may see evidence of its activity in the first few seconds as lines of the waterfall display appear to skew to the left or right before settling down to a perfectly vertical orientation.

Now click the RIG CONTROL/PTT tab and fill in the COM port number for the line you are using to key your transceiver between transmit and receive. You can test this function by clicking on the TEST PTT button. When you do, your transceiver should quickly bounce into the transmit mode and then back to receive.

Below this section, you'll see a row labeled *Ham Radio Deluxe*, *OmniRig* and *Commander*. This is only important if you are using one of these three pieces of software to control your transceiver. In most instances you can ignore this row.

Next, click on the tab labeled HEARD LIST/RB STATISTICS. If your station has an Internet connection, *JT65-HF* will share information about the stations you are hearing so that other amateurs can observe propagation conditions or test their equipment. Sharing this information is harmless and very helpful to the ham community at large. Look at the reports on **http://jt65.w6cqz.org/receptions. php**. I'd suggest that you join the party by filling in your call sign and a short

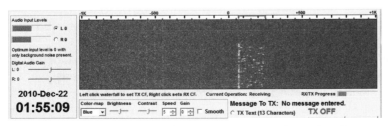

Configuration

Station Setup | RB/PSK Reporter/Rig Control | Heard List/RB Statistics | Macros | Si570 USB Control | Colors | Diagnostics

[Optional] Call for PSK Reporter or RB. Suffixed/Prefixed is allowed here. DO NOT add -1 -2 etc.   Rig Control Reported QRG Hz

WB8IMY   10130000

☑ Spot Receptions via RB Network   ☑ Spot Receptions via PSK Reporter   ☐ Operate RB in Offline Mode

[Optional] Antenna Description for PSK Reporter.

33 foot vertical antenna

OmniRig Settings
☑ Use OmniRig ⦿ Radio 1 ○ Radio 2

Ham Radio Deluxe Settings
☐ Use HRD ⦿ Instance 1 ○ Instance 2

DX Labs Commander Settings
☐ Use Commander

Save Settings and Close Window

If you want to contribute reports to the reverse beacon networks, just fill in your call sign and put check marks in both SPOT RECEPTIONS boxes. You don't need to specify your rig control software unless you are actually using software to read your transceiver's frequency.

description of your antenna system.

On this page you'll see a row labeled *OmniRig, Ham Radio Deluxe* and *Commander*. *JT65-HF* has the ability to read the operating frequency from your radio through one of these programs. You do **not** need this function to enjoy *JT65-HF*; it is more of a convenience feature, as you'll see later. If you are using one of these applications for rig control, go ahead and click "enable" in the appropriate box. Otherwise, leave it alone.

All the other tabs are safe to ignore at this time. Their functions allow you to customize *JT65-HF* and to keep things simple we're going to stick with all the default settings for now. Click SAVE SETTINGS AND CLOSE WINDOW.

The JT65-HF Main Screen

Let's take a look at the *JT65-HF* main screen, section by section, starting at the top. See **Figure 4.1**.

The waterfall display dominates most of the top portion of the *JT65-HF* main screen. Whenever *JT65-HF* is running, it sweeps through your receive audio spectrum from 0 to 2000 Hz. Every signal it detects appears in this window.

You'll notice that the waterfall is divided into two halves to the right and left of the center "zero" point. The display markers are positive to

Figure 4.1 – The top portion of the *JT65-HF* screen, including the audio input level controls on the left-hand side.

the right of the zero (0 to 1000) and negative to the left of the zero (0 to -1000). Along the top of the waterfall you'll see a red bracket. If you click your mouse cursor within the waterfall the red bracket will move to the position you just clicked. The bracket represents your 200 Hz transmit/receive window.

JT65-HF can operate in simplex (transmitting and receiving on the same frequency) and split (transmitting and receiving on different frequencies). Most of your contacts will be simplex, but it is worthwhile to know that *JT65-HF* has split-frequency capability. Two brackets appear when operating split – red for the transmit frequency and green for the receive frequency.

On the left side of the waterfall, you have the right and left channel audio input controls. When the band is quiet (when there are no signals), you should adjust the right and left channel controls to achieve 0 dB on both channels. Too much or too little audio makes it difficult to decode signals. As Goldilocks observed in the children's tale, the best porridge was the one that was not too hot and not too cold – just right.

Below the waterfall you'll see controls for color, brightness, contrast, speed and gain. You're safe leaving all these at their default settings. There is little need to change them unless you're having difficulty viewing the waterfall, or unless you are running *JT65-HF* on a slow computer.

Moving to the lower right section we have a number of checkboxes and buttons. These may seem confusing but their functions will become more apparent when we make our first contacts.

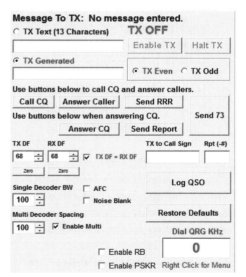

The lower-right portion of the JT65-HF screen.

MESSAGE TO TX is the text you are sending to the other station. You can send text manually in the TX TEXT window, but you are limited to 13 characters. Immediately below this window you see the red-labeled TX GENERATED window. This window is for transmitted text that *JT65-HF* generates automatically, either when you click on one of the buttons below, or when you click on a line in the decoding window. Again, the function of this window will become clear as we step through your first contact.

To the right of the message windows are buttons to enable or halt transmission. Below these buttons is the section that allows you to choose whether you wish to transmit on even or odd minutes. Remember that with JT65 you transmit and receive in turns – during one minute you transmit and during another minute you receive.

Therefore, one minute will be an odd-numbered minute (such as 2105 UTC) and the next minute will be an even-numbered minute (such as 2106 UTC).

In most instances *JT65-HF* will make the choice for you. However, when calling CQ you get to choose when you will begin transmitting – on an even or odd minute. You definitely don't need to worry about even and odd minutes when you are answering someone else's call. That station has already selected which minute (even or odd) he will use. When you respond, *JT65-HF* will automatically choose the opposite minute.

Below the text-generating buttons you'll find several other interesting sections. The TX DF and RX DF sections are for split-frequency applications. For the vast majority of your JT65 contacts you will work simplex, so you want to leave the TX DF = RX DF box checked. An exception to the rule is when you are calling CQ, which we'll address later.

Below the TX DF and RX DF sections you'll find the controls for SINGLE DECODER BW (Bandwidth) and MULTI DECODER SPACING. This is another set of controls you'll rarely have a reason to change from their default values. Normally you'd never disable the multiple-signal decoder unless you are using a particularly slow computer. The default of 100 Hz for single bandwidth is usually adequate.

Put a checkmark in the AFC (Automatic Frequency Control) box so that *JT65-HF* can compensate for stations that may be drifting a bit. If you live in an area with frequent storms or other sources of noise, put a checkmark in the NOISE BLANK box as well.

ENABLE MULTI decoder does exactly that. When enabled the decoder will attempt to find and decode all possible JT65 signals within the 2 kHz passband. Unless you are using a slow computer that has difficulty processing so much information at once, always leave this box checked.

To the right you'll find the LOG QSO button. When you click this button *JT65-HF* will save your contact information (who you worked, when, etc) to the file **jt65hf_log.adi** in the *JT65-HF* directory in a standard ADIF format that you can import into your computer logging software.

Finally, at the bottom of the lower right portion of the *JT65-HF* window you'll find two checkboxes labeled ENABLE RB and ENABLE PSKR, along with a sizeable window labeled DIAL QRG KHZ. If you are using transceiver-control software to read the transceiver frequency, the frequency will appear in this window. If you're not, you can right click your mouse cursor in this window and select from the list of common JT65 frequencies.

Why would you bother showing your operating frequency in the QRG window? The answer is that if you've checked the ENABLE RB and ENABLE PSKR windows *JT65-HF* will automatically access your home Internet connection and share your data (the stations you've heard and how strong they were) with W6CQZ's reverse beacon (RB) website and with the PSKReporter (PSKR) website – if you've enabled this feature in the station setup screen. The information is extremely helpful to your fellow amateurs who study propagation, experiment with new antennas and so on. To make the information useable, however, they need to know your listening frequency; that's what the QRG window is all about.

This is also why you normally include your grid square in your first JT65 exchanges. The reverse beacon systems use grid squares to calculate the approximate distance between stations.

Watch the Action

As is the case so often in Amateur Radio, when trying a new operating mode the best practice is to spend plenty of time listening *first*.

Start by looking at the list of JT65 frequencies in **Table 4.1**. Select a frequency and place your transceiver in the USB mode. Tune to that frequency and start the *JT65-HF* software. If there are stations transmitting, you'll see them in the waterfall display right away. You may notice that their signal traces seem to curve somewhat before settling into a regular pattern. That's a symptom of *JT65-HF* running its sample correction routine on your sound device.

When everyone stops transmitting at about the 48-second mark, use the opportunity to quickly adjust the audio gain controls to achieve 0 dB on both channels.

As the next minute begins, you should hear or see other stations starting their transmissions. Just sit back and watch. You'll notice that each transmission is comprised of a line – possibly a broken line – with several dots appearing to the immediate right. The line represents the synchronizing tone and the dots are all the remaining tones.

When the clock reaches the 48-second point, everyone should stop transmitting automatically. Within the next few seconds you should see text appear in the window at the lower left.

Some of the text may be highlighted in green. These are stations calling CQ. You'll see other text shaded in gray. These are the actual exchanges.

Look at the sample decoding screen in **Figure 4.2** and I'll explain what you are seeing. First, let's decipher the headings along the top of the window, beginning at the far left.

UTC: The time the signal was decoded in UTC.

Table 4.1
Common JT65 Frequencies

(All frequencies assume a transceiver display in USB mode.)

1838 kHz
3576 kHz
7076 kHz (European stations
　　　　often use 7039 kHz)
14076 kHz
10139 kHz
18102 kHz
21076 kHz
24920 kHz
28076 kHz

Double click an entry in list to begin a QSO. Right click copies to clipboard.

UTC	Sync	dB	DT	DF	Exchange
16:22	7	-11	-1.3	466	B TF3CY WA0SSN EN34
16:22	2	-8	-1.0	153	B AJ1E PA4C RRR
16:22	3	-8	-1.1	-377	B CQ EA1YV IN52
16:22	3	-17	-0.9	-931	B CQ PG1A JO21
16:21	6	-16	-0.8	692	B CQ ON4LBN JO20
16:21	6	-9	-0.4	466	B F5GVA TF3CY -08
16:21	4	-14	0.4	156	B PA4C AJ1E R-08
16:20	5	-10	-1.4	466	B TF3CY WA0SSN EN34
16:20	6	-7	-0.9	156	B AJ1E PA4C -08
16:20	7	-8	-0.7	-380	B WB0SOK EA1YV 73
16:20	6	-18	-0.9	-931	B CQ PG1A JO21

| Clear Decodes | Raw Decoder | Station Setup | Decode Again |

Figure 4.2 – The *JT65-HF* signal decoding window.

Sync: This is a measurement of the strength of the synchronizing tone. The higher the number, the better the sync signal.

dB: The strength of the JT65 signal in decibels. The lower the number, the stronger the signal. Zero dB is the strongest possible.

DT: How much the decoded station's time deviated from your time, measured in seconds or fractions of seconds. Ideally the decoded stations should be within two seconds of your computer's time, preferably less than one second. Even so, I've seen *JT65-HF* successfully decode stations whose signals deviated by as much as eight seconds.

DF: How far the signal frequency deviates, in Hertz, above or below the zero center point of the waterfall display. A negative number is a signal to the left of zero; a positive number is a signal to the right of zero.

Exchange: The text the transmitting station actually sent. If you see two call signs, the transmitting station is the *second* call sign.

Just to the left of the exchange text you'll see either a **B** or a **K**. This is a reference to the kind of error correction algorithm that *JT65-HF* used to validate the text. B stands for *BM*, a simple Reed Solomon algorithm. K means *KVASD*, a much more complex algorithm. One way to think about this is to imagine that a K means that *JT65-HF* had to work particularly hard to make sense of the signal. If so, this station may present a challenge if you attempt to complete a contact.

As you observe the exchanges you'll probably see a pattern emerging. JT65 exchanges usually, though not always, follow a strict sequence. It goes like this, starting at 2102 UTC …

2102   CQ WB8IMY FN31

WB8IMY has begun sending CQ on an even minute from grid square FN31.

2103   WB8IMY N1NAS EN72

N1NAS replies and tells WB8IMY that he is located in grid square EN72.

2104   N1NAS WB8IMY -11

WB8IMY replies with a signal report of -11 dB.

2105   WB8IMY N1NAS R-15

N1NAS acknowledges the signal report from WB8IMY with an "R" followed by a signal report (-15).

2106   N1NAS WB8IMY RRR

WB8IMY sends "RRR," which means "Roger, roger, roger." Everything has been received and the exchange is complete.

2107   WB8IMY N1NAS 73

N1NAS sends 73 – best wishes.

2108   N1NAS WB8IMY 73

WB8IMY sends his 73 as well. The contact has ended.

Instead of sending 73, you'll often see stations sending bits of "free hand" text instead. They are doing this by typing the text into the TX TEXT window. You may see something like **40W LOOP ANT**, which is shorthand for "I'm running 40 W to a loop antenna."

While all this may seem rather perfunctory – it certainly isn't a conversation – the exchange qualifies as a valid contact. You can use JT65 contacts for ARRL awards such as DXCC or Worked All States, for example.

Let's Answer a CQ

Once you feel comfortable monitoring JT65 activity, why not try answering a CQ? This can be substantially more exciting than it seems, especially when you realize that you have to make choices and click your mouse with just seconds to spare.

Before you begin, check your computer time calibration and make sure it is up to date. If you are

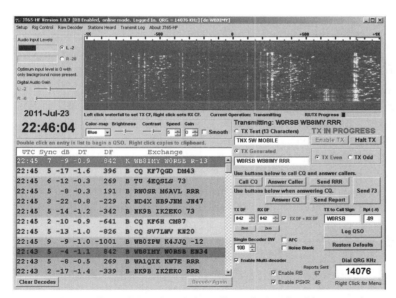

Look at the top line of the signal decoding window in this example. WØRSB is acknowledging my report (the "R") and giving me a report of −13 dB.

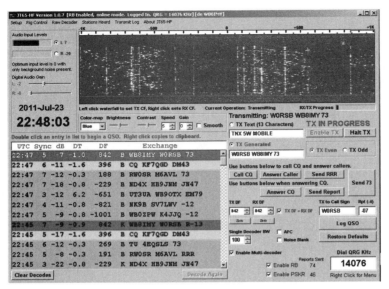

On the top line inside the signal decoding window you'll see that WØRSB is ending the contact by sending "73."

running *Windows Vista* or *Windows 7*, open CONTROL PANEL and click on DATE AND TIME. In the Date and Time window, click on the INTERNET TIME tab and then click the CHANGE SETTINGS button. In the next window select the time server (I use the NIST server) and click UPDATE NOW. You don't necessarily need to do this before every *JT65-HF* operating session, but it doesn't hurt. If you're running *Windows XP*, you may want to consider the free time-synchronizing application *Dimension 4*, which you can download at WWW.THINKMAN.COM/DIMENSION4/. Install the application and set it up so that it loads and runs constantly in the background whenever you start your computer. By the time you read this book I suspect a similar application will be available for *Windows 7* as well.

Finally, reduce your transceiver's RF power output to about 50 W; you don't need 100 W to make JT65 contacts. If you are using any sort of software to read the transceiver's frequency, start it *before* you start *JT65-HF*. If you are participating in the reverse beacon system, select your frequency by right clicking in the QRG window, or allow your transceiver control software to fill in the window automatically.

Tune to a JT65 frequency (14.076 MHz is often a good choice) and watch the traffic for a few minutes. You're looking for CQ transmissions highlighted in green.

The green lines will appear as soon as your computer decodes the signals, usually at about the 48- to 52-second point. When you see one, you have only eight seconds to double-click on the line to begin your reply. Think fast!

When you double-click on the green CQ line, you'll see that the text of your transmission appears automatically in the **TX Generated** window. Don't worry about selecting the even or odd minute; *JT65-HF* "knows" that you want to

In *Windows Vista* and *Windows 7*, you access the time settings by clicking on the DATE AND TIME icon in Control Panel.

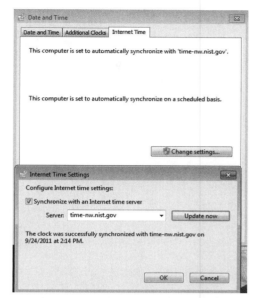

Once you're in the DATE AND TIME window, click the INTERNET TIME tab, followed by the CHANGE SETTINGS button.

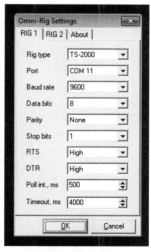

In this example *OmniRig* (www.dxatlas.com/OmniRig/) is being used to read the transceiver's frequency automatically. Whichever software you may be using for rig control (if any) must be started before you start *JT65-HF*.

transmit during the next available minute and it will select even or odd accordingly.

At the top of the minute, when the two right-hand digits read "00," *JT65-HF* will key your radio and begin transmitting. You'll see **TX IN PROGRESS** in red letters above the ENABLE TX/HALT TX buttons. This is a good time to check the ALC activity on your radio's meter. If the ALC activity is excessive, reduce the transmit audio from your interface or computer.

JT65-HF will continue transmitting your message until just before the 48-second point. Sit back and wait. At the top of the next minute, the other station will begin transmitting. If he received your transmission and replied, you will see his text, which will contain a signal report, highlighted in red after *JT65-HF* decodes it. The line is shaded in red for a good reason: You need to take action immediately!

Quickly double-click on the red line and *JT65-HF* will fill in the text for your next transmission. You'll send an "R" to acknowledge his report followed by a report of your own. *JT65-*

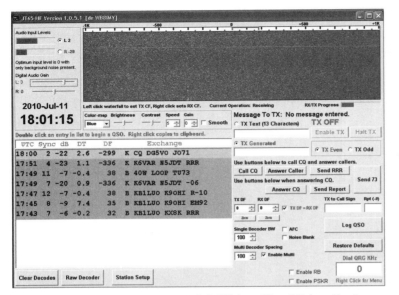

In this *JT65-HF* screen, notice that DG5VO is calling CQ (see the top line in the decoding window). If you double-click your mouse on this line, *JT65-HF* will automatically create the proper text and set you up to transmit on the next opposite (even or odd) minute.

HF will measure the other station's signal strength and will insert the number for you. When the clock reaches the top of the minute, *JT65-HF* will begin sending.

If all goes well, his next transmission will be "RRR" highlighted in red. Double-click on this line and *JT65-HF* will finish the contact by sending 73.

You Answered the CQ, but He Didn't Reply. Why?

It is possible that the other station simply didn't hear you. If you see another green CQ line from the station, double-click on the line and try again.

It is also quite possible that he responded to someone else. If you see his call sign in a gray shaded line, he is responding to another station. Quickly click on the HALT TX button so that you won't transmit again and interfere with the contact.

Did He Send a Free Hand Text?

At the end of your contact the other station may send you a brief message, such as a description of his antenna or how much RF output he is using. This text won't contain his call sign, so *JT65-HF* won't flag it with a color other than gray.

So if you see a message, how do you know it was from him?

Look at the DF (frequency deviation) portion of the decoded line. Does the deviation figure match the one that appeared during your exchanges (the text highlighted in red)? If so, that text was from your partner and was definitely meant for you.

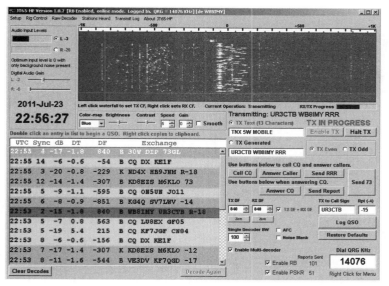

Call CQ

When you've become comfortable answering a few JT65 CQs, it is time to try one of your own.

While it is almost always best to have TX DF = RX DF enabled in *JT65-HF*, this is the one instance when it is exactly what you should *not* do. The best way to explain this is with a hypothetical example.

Let's say you've picked a clear spot in the waterfall display and you clicked on that spot to position the red bracket accordingly. You click the CALL CQ button and *JT65-HF* responds by generating the appropriate text and setting you up to transmit on the next odd or even minute, depending on which one you selected.

Look at the top line in the *JT65-HF* signal decoding window. UR3CTB just sent "30W DP 73GL" as free hand text, which translates to "I'm running 30 W to a dipole antenna. 73 and good luck." How do I know this came from UR3CTB? Examine the line about halfway down the window, the one where UR3CTB is giving me a signal report of -18. Notice that the DF (Deviation Frequency) is 840 – the same DF as the line with the free hand text. It's safe to say that the message is from UR3CTB.

You transmit and receive a reply. You double-click on the reply but you've left the TX DF = RX DF function on and the station calling you is at some offset (different frequency) from your transmitting frequency. What happens? *JT65-HF* automatically adjusts your transmit frequency to match his. The other operator replies by double-clicking on your decoded text, which means that his bracket will shift to match your new transmit frequency. When his station transmits you'll double-click on his text and your bracket will move again because his transmit frequency has changed. Back and forth you go as both of you "walk" through the waterfall changing frequencies with every transmission and causing untold interference. This is not good.

The cure is to uncheck the TX DF = RX DF box before you begin sending your CQ. This will keep your frequency stable under the red bracket while the program shifts to whatever other frequency is necessary to decode the answering

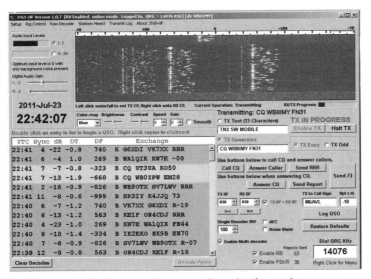

In this example, I chose my frequency by selecting a clear space at the far right-hand side of the waterfall. Next I clicked the CALL CQ button, which automatically inserted the CQ text in the TX GENERATED window. Note the TX IN PROGRESS message above the ENABLE/HALT TX

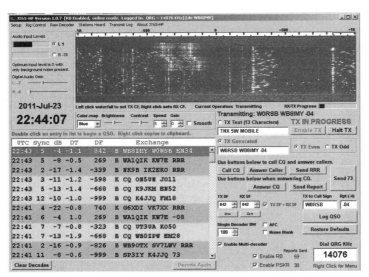

See the top line in the signal decoding window. WØRSB has answered my CQ from his location in grid square EN34.

station. It will label this frequency with a green bracket.

When someone replies to your CQ, you'll see the text highlighted in red as before. Just double-click on the red lines and *JT65-HF* will create the proper exchanges automatically. The trick is to stay on your toes and click on the appropriate line before the next minute begins. Otherwise, *JT65-HF* will simply re-send the previous line. The other station will be confused and you'll both have to wait through another set of exchanges to complete the contact.

The Joy of JT65

JT65 on the HF bands has been a boon for amateurs living under the burdens of antenna or power restrictions. At the time of this writing, there are amateurs who've contacted dozens of countries with just a few watts and indoor antennas. In **Figure 4.3** you'll see a 20-meter dipole antenna that I made by mounting two Hamstick-style mobile antennas back to back. This is a particularly poor

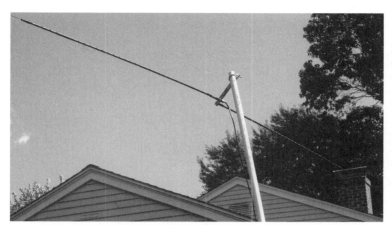
Figure 4.3 – A 20-meter dipole made from two Hamstick-style mobile antennas joined together.

antenna and yet I was able to use JT65 to make contact with stations on the opposite side of the world with just 5 W of RF.

The other interesting aspect of JT65 is that you can operate in a leisurely fashion throughout most of the contact. You can carry on a conversation with a loved one while occasionally glancing at the *JT65-HF* decoding window. The only time you need to act quickly is in that 8- to 10-second span between the time *JT65-HF* decodes a transmission and the beginning of the next minute. Otherwise, you can kick back and relax.

Chapter Five

MFSK and Olivia

The Short Scoop
These two modes are somewhat
uncommon, but both allow you to enjoy
conversations under just about the worst
conditions imaginable.

MFSK

MFSK – Multi-Frequency Shift Keying — is really a type of super-RTTY. Instead of using just two tones as RTTY does, MFSK uses many more. Sixteen-tone MFSK (MFSK16) is the most common.

The MFSK technique was developed during the heyday of teleprinter HF communications as a way to combat multipath propagation problems, providing reliable point-to-point communications with relatively simple equipment. Piccolo, for example, is a similar mode used on diplomatic links between England and Singapore "back in the day," and it typically provided good copy for an hour after the RTTY links had faded out. The technology then was electromechanical, but several very important principles were recognized at the time…

• The performance (reduced error rate) improved as the number of tones used increased.

• The performance was best when the least number of symbols was used to represent each transmitted text element.

• With a special integrating detector, tones as closely spaced as the baud rate could be uniquely detected without cross-talk.

Piccolo used two symbols per text character—compare this with 7.5 symbols per character for RTTY and anywhere from three to 12 for PSK31. MFSK16 uses only *one* symbol per signaling element. With MFSK modes the baud rate (rate at which symbols are transmitted) is rather lower than the text rate. This is because each symbol carries more information in its frequency properties than RTTY or PSK. While this may seem confusing, the technique has the advantage that the longer symbols are easier to detect in noise, have a narrower bandwidth, and are much less affected by multipath errors.

Piccolo originally used as many as 32 tones, but the most common form uses six. MFSK has been tested with as many as 64 tones; the weak signal variant MFSK8, uses 32.

Anatomy of an MFSK Signal

What does an MFSK signal consist of? Well, the most common form uses 16 tones, sent one at a time at 15.625 baud, and they are spaced only 15.625 Hz apart. To put this tone spacing in perspective, recall from Chapter 2 that a "narrow" PSK31 signal is more than 31 Hz wide—twice as wide as the space between just two MFSK16 tones.

Each tone represents four binary bits of data. The whole transmission is 316 Hz wide, which is a bit narrower than a RTTY signal. And like a RTTY signal, an MFSK16 signal easily fits within the passband of a 500 Hz CW filter. In fact, if you're suffering from interference you can use the same IF filters for MFSK16 that you'd use with RTTY.

When you hear an MFSK signal for the first time, you'll notice its peculiar musical quality. Some compare it to the sound of an old-fashioned carnival calliope.

Like RTTY, the signal is constant amplitude—it does not require a linear transmitter to maintain a clean signal. Driving the transmitter too hard on MFSK16 will *not* make the signal any wider. However, keep in mind that the MFSK16 duty cycle, like RTTY, is high. This can be hard on your transceiver if you are running at full RF output.

To ensure that text is received with an absolute minimum of errors, MFSK utilizes an excellent Forward Error Correction (FEC) technique, using Viterbi decoder routines by Phil Karn, KA9Q, and a clever self-synchronizing interleaver developed for MFSK by Nino Porcino, IZ8BLY. The typing rate, even with FEC, is over 40 WPM. As complicated and arcane as this may sound, the bottom line is that MFSK16 can be successfully decoded under truly dreadful conditions.

On the Air with MFSK

If you are running *Windows XP* there is a piece of dedicated MFSK software by IZ8BLY known as *Stream*. You can download it free at **http://antoninoporcino.xoom.it/Stream/**. Unfortunately, *Stream* does not work well with *Windows Vista* or *7*.

You'll otherwise find MFSK among the selections in the multimode software packages we discussed in Chapter 1. Some of these applications will offer several MFSK modes from MFSK4 all the way to MFSK64. Regardless of the software, MFSK will run perfectly with any voice transceiver using the same sound device interface you'd use for PSK31 or other HF digital modes. One of the advantages of setting up an HF digital station around sound-device software is that switching from one mode to another is as easy as clicking your mouse. There are no hardware changes to worry about.

MFSK16 is most often heard just above the same frequencies used by PSK31 (see Chapter 2). Twenty meters is the most popular band by far. If you opt to try 20 meters first, listen between about 14.072 and 14.076 MHz with your transceiver in the USB mode. If you'd like to hear what MFSK16 sounds like, go to the *Get On the Air With HF Digital* website at **www.arrl.org/HF-Digital** and listen to the audio sample.

Depending on the software you are using, you may discover that tuning an MFSK16 signal takes a certain amount of patience. Because the tones are closely spaced and the software decoding filters very narrow, you must have a stable transceiver and you must use the tuning provided with the software (in the waterfall display). In the waterfall you'll notice a line that occasionally appears along the left-hand side of the column of dots that make up the MFSK16 signal. This is the *idle carrier*, the lowest of the 16 tones. This carrier is transmitted briefly at the start and end of each transmission, or whenever the operator stops typing. If your software uses a "tuning bracket" in the waterfall, you need to move the left-hand edge of this

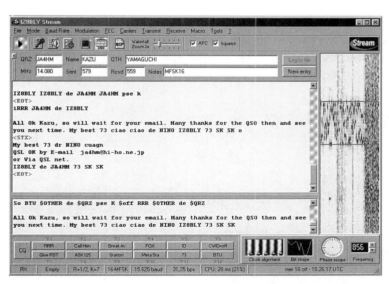

Nino Porcino, IZ8BLY, developed the *Stream* software for MFSK. You can download it free at http://antoninoporcino.xoom.it/Stream/. Unfortunately, *Stream* does not work well with *Windows Vista* or *7*. It only works reliably with *Windows XP*.

bracket until it is on top of the idle carrier line.

Each time you nudge the bracket, wait for several seconds to give the software time to synchronize and begin decoding. If the bracket isn't in the correct position, you'll see gibberish text or nothing at all.

Once you've found the right spot, almost perfect text will start to appear on the screen, although it is delayed by some 3-4 seconds as the data trickles through the error correction system and appears one or two words at a time.

If you have found someone calling CQ, you can answer them easily. Just wait until they stop transmitting and then click on the button your software uses to key your transceiver. Your interface should send the MFSK16 audio to your radio and away you go!

Leisurely type in the transmit buffer window and then click the appropriate software button to return to receive. When the other station transmits you will notice that a few seconds may pass before text begins to appear. This is normal.

All multimode software includes macros that you can certainly use in your MFSK16 conversations, but the tradition is to carry on the chats "manually" without the use of macros.

Performance

For QSOs using short path, say at distances to 8000 miles on 20 meters, MFSK16 works fine, but you may find PSK31 easier to use. If you are interested in low-power QRP operating, there is not much to choose between the two. On long path, over the poles, and in really difficult conditions where interference or instability is the major problem, MFSK16 is in a class by itself. It just keeps going, giving almost perfect copy when the signal is barely audible. High power isn't necessary.

MFSK16 is one of the best conversational modes yet for digital work on the lower bands. If you are into traffic handling, or sending bulletins on 80 or 40 meters, give this mode a try. MFSK16 will work over thousands of miles on 80 meters at night, running one watt and with 90% perfect copy. That's why the ARRL's

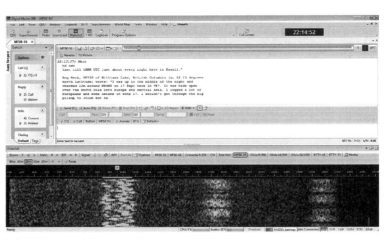

W1AW transmitting a bulletin with MFSK16.

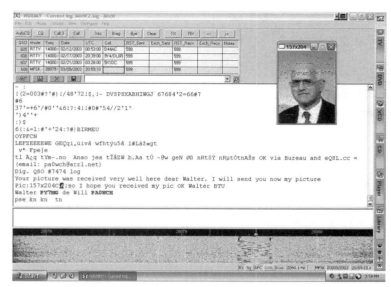

You can even send tiny pictures with MFSK16 if your software supports this function.

Headquarters station W1AW includes MFSK16 among its digital modes for bulletin transmissions.

Although not noticeably better on the low bands, MFSK8 is great to have when the band starts to die. It is definitely more sensitive than MFSK16, and although tuning is very tight, and typing speed is down to 25 WPM, it will allow you to complete the conversation with almost perfect copy.

With MFSK16 it may also be possible to swap more than text. Some multimode programs will allow you to send tiny *images* within an MFSK16 conversation. For example, *MixW* for *Windows* software supports this image function very smoothly. If you're chatting in MFSK16 and the other station begins sending an image, *MixW* will automatically open a small auxiliary window and display the picture. If you're a *MixW* user, you can also create a macro to send an image of your own. Just right click your mouse on one of the *MixW* macro buttons that you rarely, if ever, use and clear out whatever text is there. Now insert the following text…

<PIC?N%C>

Give the button a label you'll recognize such as "MFSK Pix." When you are transmitting in MFSK16, click this button and you'll be prompted to select an image file from the *MixW* directory on your hard drive (put some favorite images there for times like these). Just click the file name, then OPEN. *MixW* will immediately begin sending the image to the other station.

Olivia

Olivia was the brainchild of Pawel Jalocha, SP9VRC, and it was named after his daughter *Olivia*. It is one of the best conversational digital modes in terms of its ability to print readable text in poor conditions. The signal can be decoded

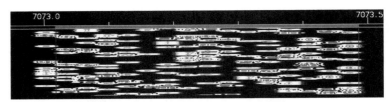

This is what a 16-tone, 1000 Hz bandwidth Olivia signal looks like in your waterfall display.

even when it is 10 to 14 dB below the noise floor (i.e. when the amplitude of the noise is slightly over three times that of the signal). It can also decode well under noise, fading and interference. The only HF digital mode that can out-perform Olivia is JT65, but you can't have a conversation with JT65!

Olivia has many formats, some of which are considered standard, and they all have different characteristics. The formats vary in bandwidth (125, 250, 500, 1000 and 2000 Hz) and number of tones used (2, 4, 8, 16, 32, 64, 128 or 256). This makes it possible to have *40* different Olivia formats that have different characteristics, speeds and capabilities. Luckily only a few are commonly used.

You'll find Olivia in multimode software such as *MixW*, *MultiPSK*, *Fldigi* and *Ham Radio Deluxe*. At the time this book went to press, Olivia was *not* offered in the two most popular Mac OS programs: *cocoaModem* and *MultiMode*.

The standard Olivia formats (bandwidth/tones) are 125/4, 250/8, 500/16, 1000/32 and 2000/64. However, the most commonly used formats in order of use are 500/16, 500/8, 1000/32, 250/8 and 1000/16. After getting used to the sound and look of Olivia in the waterfall, though, it becomes easier to identify the format when you encounter it. About 98% of all current Olivia activity on the air is one of the seven following configurations: 1000/32, 1000/16, 500/16, 500/8, 250/8, 250/4 and 125/4.

Common Olivia frequencies are shown in **Table 5.1**. Olivia announces itself with a musical sound not unlike MFSK16, but it will appear to be wider in the software waterfall display. You can listen to the sound of a typical Olivia signal at the *Get On the Air with HF Digital* web page at **www.arrl.org/HF-Digital**.

On the Air with Olivia

Place your transceiver in the Upper Sideband (USB) mode and start tuning around. If you suspect you've found an Olivia signal, observe its appearance in your waterfall display. Choose an Olivia mode from your software MODES menu that results in a tuning bar (or bracket) that appears to match the width of the Olivia signal in the waterfall. Position the bar on the signal and wait patiently. If you thought MFSK16 was slow to decode and print, Olivia can be much slower. That's one of the tradeoffs Olivia makes for its amazing performance.

If you don't see coherent text after 15 seconds or so, you may have chosen the right bandwidth but the wrong number of tones. Quickly go back to the

Table 5.1

Common Olivia Frequencies

All frequencies are Upper Sideband (USB).

Band (Meters)	Frequency (kHz)
160	1835 to 1838
80	3583.25 and 3577
40	7035 to 7038
30	10141 to 10144
20	14072 to 14075.65 and 14106.5
17	18102.65
15, 12, 10	1-2 kHz above PSK31 activity: 21072, 24922, 28122

MODES menu and try again with a different number of tones in the same bandwidth.

An experienced Olivia operator knows that this initial decoding process can take a while, so he will usually send a very long CQ. An Olivia CQ lasting more than a minute is not uncommon. The idea is to give listeners plenty of time to lock in the signal.

With that in mind, it is a good idea to create a long CQ macro for Olivia that you can trigger with a single click of your mouse. It certainly beats having to manually type a very long CQ each time!

If you software supports it, try creating an "Auto CQ" macro. This macro sends a long CQ, pauses for about a minute then sends again. The Auto CQ function will allow you to putter about your station while you wait for someone to tune you in and answer. Since Olivia is not as common on the air as PSK31, this may take some time.

Once you've established a conversation, sit back, relax and enjoy. Olivia moves at a slow pace, printing only a few words on your screen at a time. This results in conversations that may last as much as an hour, perhaps even longer. Just watch the text in the receive window and type your responses in the transmit window. You don't have to wait for the other station to stop before typing your responses. On the contrary, you can enter your comments while the other station is still transmitting. When you enter the transmit mode, they'll be sent automatically.

Olivia Hints

There are a few things you can do to improve how well Olivia works at your station. Some of these tips may not necessarily be supported in the software you are using, but chances are that most will.

• **If your multimode application has a software squelch function, set it to "low" or "off."** Yes, this setting will produce some random "garbage" text on your screen, but it will also allow you to read text from very weak Olivia signals that would otherwise be invisible.

• **Exercise CQ patience.** We've touched on this already, but it bears repeating. When you call CQ on this mode be patient and wait at least

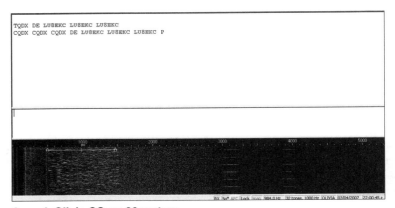

```
TQDX DE LU8EKC LU8EKC LU8EKC
CQDX CQDX CQDX DE LU8EKC LU8EKC LU8EKC P
```

A weak Olivia CQ on 80 meters.

45-60 seconds before you call again. Remember that the listening station may be effectively 5-20 seconds "behind" you, still decoding your signal after you've stopped transmitting. If you instantly hit the CQ macro button again, the other fellow may still be decoding and won't be able to respond. Wait so that he has plenty of time to decode all your text and begin transmitting his reply.

• **Don't get carried away with your Sync Integration Period.** This software setting (some programs may label it differently) determines how "deep" the software digs into the data to decode an Olivia signal. You may say, "Set it as high as possible for maximum performance," but this would be a mistake. If you set this parameter sky high, you may be waiting *minutes* before Olivia begins decoding. This is clearly absurd. If you are in doubt, go with whatever the software default setting happens to be. Feel free to experiment, though. You'll eventually find that you can indeed squeeze out better performance at higher settings. It is just a matter of how much delay you care to tolerate.

• **Keep your "Search" setting at about 8**. The Search setting (or Tune Margin) determines how far from the center frequency (in the middle of the tuning bar) the software will go to grab the Olivia signal. Once again, if you are in doubt, use the default value.

Chapter Six

PACTOR

The Short Scoop
All the modes discussed in this book
so far use sound devices as modems.
This one is different – and it essentially
guarantees error-free communication.

Hans-Peter Helfert, DL6MAA, and Ulrich Strate, DF4KV, developed PACTOR in 1991. It remains to this day the most popular of the HF *burst* modes. There are now four versions of PACTOR: PACTOR I, PACTOR II, PACTOR III and PACTOR IV. PACTOR I is considered obsolete and is little used. PACTOR II and III are in current use, primarily to connect to automated stations such as the Winlink 2000 network, which we'll discuss later.

There is no sound-device software method for sending and receiving PACTOR. As I mentioned in Chapter 1, the only way to get on the air with PACTOR is to use a stand-alone hardware *controller* (also referred to as a Multimode Communications Processor) that connects between your computer and an SSB transceiver.

We call PACTOR a burst mode because of the way it sends information. Rather than sending a continuous stream of data like RTTY, PSK31, MFSK16 or Olivia, PACTOR transmits bursts of information that take the form of short data blocks. When the data is received intact, the receiving station sends an ACK signal (for acknowledgment). If the data contains errors, a NAK is sent (for nonacknowledgment). In simple terms, ACK means, "I've received the last group of characters okay. Send the next group." NAK means, "There are errors in the last group of characters; send them again." This back-and-forth data conversation sounds like crickets chirping.

In the case of PACTOR, the long chirp is the data and the short chirp is the ACK or NAK.

Memory ARQ

In less sophisticated burst modes such as AMTOR or packet, a data block must be repeated over and over if that's what it takes to deliver the information error-free. This results in slow communication, especially when conditions are poor.

PACTOR handles the challenge of "repairing" errors in an interesting way. As stated above, each data block is sent and acknowledged with an ACK signal if it's received intact. If signal fading or interference destroys some of the data, a NAK is sent and the block is repeated. Nothing new so far—packet and AMTOR behave much the same way. The big difference, however, involves memory.

When a PACTOR controller receives a mangled character block, it analyzes the parts and temporarily memorizes whatever information appears to be error-free. If the block is shot full of holes on the next transmission as well, the controller quickly compares the new data fragments with what it has memorized. It fills the gaps as much as possible and then, if necessary, asks for another repeat. Eventually, the controller gathers enough fragments to construct the entire block (see **Figure 6.1**). PACTOR's memory Automatic Repeat reQuest feature dramatically reduces the need to make repeat transmissions of damaged data. This translates into much higher throughput.

PACTOR has the capability to communicate at varying speeds according to band conditions. PACTOR throughput is enhanced by using *Huffman coding* that reduces the average character length for improved efficiency.

What Do You Need to Run PACTOR?

Assembling a PACTOR station is very simple. All you need are the following:

• **An SSB transceiver.** The output of the PACTOR controller is audio, which can be applied at the microphone or accessory jack. The controller also operates the transmit/receive keying line. The only catch is that the transceiver must be capable of switching between transmit and receive in less than about 20 milliseconds. Most modern transceivers meet this requirement. If you're unsure, check the *QST* magazine Product Reviews. The ARRL Lab routinely tests this function in every transceiver review.

• **A controller with PACTOR capability.** Current controllers by Timewave and older Kantronics and AEA controllers support PACTOR I, but only controllers manufactured by Special Communications Systems (SCS) offer PACTOR II, III and IV.

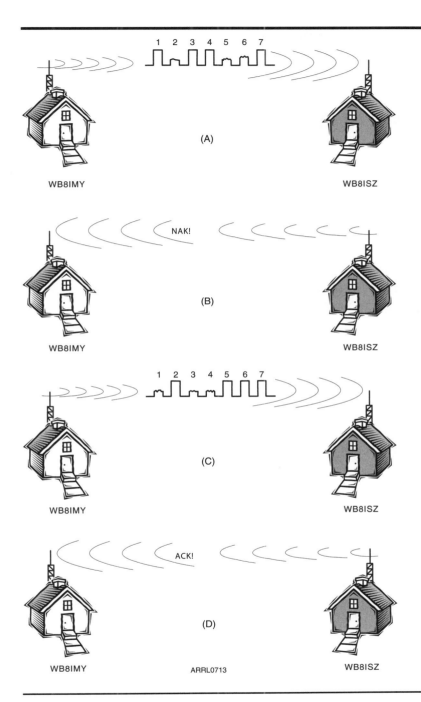

Figure 6.1 – PACTOR Memory ARQ at work. In this example, WB8IMY sends a burst of data to WB8ISZ (A), but bits 2, 5 and 6 are corrupted due to noise, fading or interference along the way. The controller at WB8ISZ memorizes the valid data along with the positions of the corrupted bits. It sends a NAK (B) to demand a repeat transmission. On the next transmission (C), bits 1, 3 and 4 are corrupted, but that's not a problem. Why not? The answer is that WB8ISZ's controller already has valid bits for 1, 3 and 4–it doesn't matter if these are corrupted. Instead, on the second transmission, bits 2, 5 and 6 made it through intact. WB8ISZ now has a complete, error-free data frame so it sends an ACK (D) telling WB8IMY that all is well and requests the next burst of data. This technique greatly reduces the number of transmissions needed to communicate error-free when conditions are poor.

ARRL0713

• **A computer running the controller's companion software.** Most compatible software is written for Microsoft *Windows*.

When assembling your PACTOR station, use the guidelines discussed in Chapter 1. The connection to the computer is made with a serial cable in older controllers, so you may need a USB/serial adapter if your computer doesn't have a serial port. The latest SCS PACTOR units offer a USB connection.

CQ PACTOR

Although most PACTOR activity involves connecting to automated systems, you can still occasionally enjoy a "live" conversation. You'll tend to find these in the same frequency segments used for RTTY (see Chapter 3).

You call CQ on PACTOR using *forward error correction*, or FEC. The FEC signal sounds like very fast RTTY, but in reality it is a stream of data in which each character is repeated twice for redundancy—there are no ACKs or NAKs. Obviously, this means that an FEC transmission is not error-free, but the copy is good enough to pull out the call sign of the sending station. When you're sending a PACTOR CQ, that's really all that matters.

If you hear a signal that you suspect is a PACTOR FEC CQ, tune carefully until your controller indicates that it has locked (synchronized) with the FEC signal. Within a short time, you should begin to see text on your screen.

CQ CQ CQ DE WB8IMY WB8IMY WB8IMY
CQ CQ CQ DE WB8IMY WB8IMY WB8IMY
CQ CQ CQ DE WB8IMY WB8IMY WB8IMY K K

Notice that this CQ uses several short lines of text rather than a few long lines. This helps stations synchronize more easily.

Starting a Conversation

Answering a CQ in PACTOR is straightforward. Depending on the software you're using, it may be as simple as entering:

CALL N1BKE
or, **CONNECT N1BKE**

at the **cmd:** prompt. Some types of software streamline the process even further. There may be pop-up boxes where you simply enter a call sign.

Once you make contact, the conversation proceeds in turns. This means that one station "talks" (the ISS or information sending station) while the other "listens" (the IRS or information receiving station). When the ISS has had his say, he sends a special control signal known as the *over command* that immediately reverses the roles—suddenly you are the ISS and he is the IRS. Introduce yourself and ask a question about where he lives, or what he does for a living. Use the over command to flip the link again. A conversation is underway!

Depending on the software you are using with your controller, sending the over command is usually as easy as tapping a single key. The software usually provides some sort of visual indicator to show which mode you are in—IRS or ISS—in case you become confused! If you forget to send the over command, your stations will simply sit there and chirp mindlessly at each other. Fortunately, the IRS can send an over command and flip the link for you if necessary. This is known as a "forced over."

PACTOR II

You can think of PACTOR II in terms of being a supercharged version of PACTOR I. In good conditions, PACTOR II boasts a data transfer rate up to six times faster than PACTOR. At the same time, PACTOR II is capable of maintaining links in conditions where the signal-to-noise ratio is at -18 dB. This means that PACTOR II can carry on an exchange even when the signals are virtually inaudible. Such remarkable performance is also conservative when it comes to bandwidth; a PACTOR II signal only occupies about 500 Hz of spectrum.

The SCS PTC-II PRO multimode controller supports PACTOR I, II and III as well as several other HF digital modes.

The complex pi/4-DQPSK modulation system used by PACTOR II requires DSP technology and fast microprocessors. With DSP, you let the software decide the composition of the signal, not the hardware. DSP is much more flexible, allowing you to create the signal you want without having to worry about hardware filters and so on. The tradeoff is that you must use high-speed microprocessors to handle all the incoming and outgoing information at a decent rate. That's part of the reason why PACTOR is implemented by SCS in dedicated hardware controllers.

PACTOR II is also backward compatible with PACTOR I. That is, a PACTOR II operator can communicate with a PACTOR I operator and vice versa.

PACTOR II Station Requirements and Operation

The requirements for a PACTOR II station are essentially the same as for a PACTOR I station: An SSB transceiver, a computer or data terminal and a PTC-II or PTC-IIe processor.

PACTOR II also operates in nearly the same fashion as PACTOR I. In fact, the link is initially established using the PACTOR I protocol and then automatically switched to PACTOR II if both stations are using PACTOR II processors. And as with PACTOR I, you must take turns during the conversation, sending an over command to allow the other operator to send data.

PACTOR III

PACTOR III was introduced by SCS in 2001 as a firmware upgrade to their PTC line of multimode controllers. When this edition went to press, PACTOR III was being used primarily in the Winlink 2000 network, although upgraded controllers remain "backward compatible" with PACTOR I and II.

PACTOR III represents a substantial improvement with its two-dimensional orthogonal pulse shaping, advanced error control coding and efficient source coding. It provides excellent performance in poor signal conditions and achieves high throughput rates under good signal conditions. The tradeoff, however, is the fact that a PACTOR III signal occupies a 2200 Hz bandwidth, as much as a typical SSB voice transmission. For this reason, PACTOR III stations often operate well above the frequencies used for PSK31, RTTY and other modes. For example, on 20 meters you will often hear PACTOR III signals between 14,103 and 14,115 kHz.

The key benefits of PACTOR III are:

• Higher throughput compared to PACTOR II. Under average signal conditions, a speed gain factor of 3-4 is possible. Under very good conditions, PACTOR III can be as much as five times faster.

• A maximum data rate of 2700 bits per second without compression. With text compression applied, a data rate of 5200 bps is possible.

As with PACTOR II, you must purchase an SCS multimode controller to use PACTOR III. This mode is not available from any other manufacturers.

PACTOR IV

SCS unveiled PACTOR IV in early 2011 with the introduction of their model DR-7800 multimode controller. The big difference between PACTOR IV and PACTOR III is speed. The PACTOR IV protocol boasts a symbol rate of 1800 baud within a 2400 Hz bandwidth using 10 speed levels, DBPSK/DQPSK (non-coherent, spreading factor 16), BPSK-32QAM (coherent) and adaptive equalizing. All this translates into astonishing throughput that is

The SCS DR-7800 Dragon controller is the only unit that supports PACTOR IV. Unfortunately, PACTOR IV is not legal for use on Amateur Radio frequencies.

potentially more than double that of PACTOR III. Squeezing that kind of HF digital performance into a 2400 Hz bandwidth required years of painstaking work.

Unfortunately, PACTOR IV is not legal for amateur use on the HF bands. Below 28 MHz American amateurs are restricted to data modes with effective symbol rates of 300 baud or less. PACTOR IV exceeds this limit substantially. On the other hand, since Military Auxiliary Radio System (MARS) HF operations take place outside the amateur bands, they aren't hobbled by that restriction. They've been using PACTOR with the Winlink network for a number of years, so the DR-7800 and PACTOR IV could prove to be a powerful new tool for US amateurs who participate in MARS.

Winlink 2000: The HF E-Mail Connection

The Internet has become the message medium of choice for most hams. But there is a sizeable group of amateurs who often travel beyond the reach of the Internet. This group includes hams at sea, travelers in recreational vehicles (RVs), missionaries, scientists and explorers. No doubt the day will come when wireless, affordable Internet access will be available from any point on the globe. Until that day arrives, however, the Amateur Radio HF digital network is a very capable substitute!

The Evolution of Winlink 2000

Dozens of digital stations worldwide have formed a remarkably efficient Internet information exchange network, including e-mail and binary file transfers. See **Table 6.1**. Running *Winlink* 2000 software and using mostly PACTOR II or III, these facilities transfer information between HF stations and the Internet. They also share information between themselves using Internet forwarding.

The network evolved in the 1990s from the original AMTOR-based *APLink* system, authored by Vic Poor, W5SMM. APLink was a network of stations that relayed messages to and from each other and the VHF packet network. As PCs became more powerful, and as PACTOR and Clover superseded AMTOR, a new software system was needed. That need brought about the debut of *Winlink*, authored by Vic Poor, W5SSM, with additions from Peter Schultz, TY1PS. *Winlink* itself evolved with substantial enhancements courtesy of Hans Kessler, N8PGR. To bring the Internet into the picture Winlink stations needed an e-mail "agent" to interface with cyberspace. To meet that requirement Steve Waterman, K4CX enlisted the help of Jim Jennings, W5EUT and Rick Muething, KN6KB, to add *NetLink*.

Early in 2000 the system took the next evolutionary leap, becoming a full-

Table 6.1
Winlink 2000 RMS Radio Gateway Stations
(as of December 2011)
Note: On ham frequencies in the United States WINMOR stations can only use the 500 Hz bandwidth.

Call Sign	Frequency (kHz-USB)	Mode	Access
DA5UDI	7051.4	PACTOR I/II/III	Public
DA5UDI	14094.9	PACTOR I/II	Public
DA5UDI	14102.4	PACTOR I/II/III	Public
DA5UDI	14107.4	PACTOR I/II/III	Public
DA5UDI	14108.9	PACTOR I/II/III	Public
DA5UDI	14110.4	PACTOR I/II/III	Public
DBØZAV	3598.9	WINMOR 500	Public
DBØZAV	7048.9	WINMOR 500	Public
DBØZAV	7049.9	WINMOR 500	Public
DBØZAV	14092.9	WINMOR 500	Public
EA8RCT	7038.7	PACTOR II/III	Public
EA8RCT	10145.6	PACTOR II/III	Public
EA8RCT	14110.0	PACTOR II/III	Public
HB9AK	10145.9	PACTOR I/II	Public
HB9MM	3605.5	PACTOR I/II/III	Public
HB9MM	7053.0	PACTOR I/II/III	Public
HB9MM	10145.0	PACTOR I/II/III	Public
HB9MM	14109.2	PACTOR I/II/III	Public
HB9XQ	3614.5	PACTOR I/II/III	Public
HB9XQ	7055.5	PACTOR I/II/III	Public
HB9XQ	14120.5	PACTOR I/II/III	Public
HB9XQ	3619.0	WINMOR 1600	Public
HB9XQ	7058.0	WINMOR 1600	Public
HB9XQ	14117.5	WINMOR 1600	Public
HP2XBA	7104.4	PACTOR I/II/III	Public
HP2XBA	10148.2	PACTOR I/II/III	Public
HP2XBA	14109.2	PACTOR I/II/III	Public
HP2XBA	18119.0	PACTOR I/II/III	Public
HP2XBA	28130.0	PACTOR I/II/III	Public
IKØOXK	7045.0	WINMOR 1600	Public
IQ4VU	7042.5	WINMOR 1600	Public
K4XV	7101.9	PACTOR II/III	Public
K4XV	14098.7	PACTOR II/III	Public
K6CYC	7103.5	PACTOR III	Public
K6CYC	10146.2	PACTOR III	Public
K6CYC	14108.5	PACTOR III	Public
K6IXA	7076.9	PACTOR I/II	Public
K6IXA	7102.4	PACTOR I/II/III	Public
K6IXA	10122.9	PACTOR I/II	Public
K6IXA	10143.7	PACTOR III	Public
K6IXA	14064.9	PACTOR I/II	Public
K6IXA	14102.7	PACTOR III	Public
K7EK-5	3569.5	WINMOR 500	Public

Call Sign	Frequency (kHz-USB)	Mode	Access
K7EK-5	7081.1	WINMOR 500	Public
K8QJH-5	3595.0	WINMOR 500	Public
KA7CTT	7067.9	PACTOR I/II	Public
KA7CTT	7071.9	PACTOR I/II	Public
KA7CTT	7101.2	PACTOR III	Public
KB1OOQ-5	3570.7	WINMOR 500	Public
KB1OOQ-5	10130.7	WINMOR 500	Public
KB1OOQ-5	14102.4	WINMOR 500	Public
KB1TCE-5	14105.7	WINMOR 500	Public
KB3JAJ-5	24927.5	WINMOR 1600	Public
KB5HCD	3595.0	PACTOR I/II/III	Public
KB5HCD	7065.3	PACTOR I/II	Public
KB5HCD	7096.0	PACTOR I/II	Public
KB5OZE-5	3586.5	WINMOR 500	Public
KB5OZE-5	7087.5	WINMOR 500	Public
KB5OZE-5	10134.5	WINMOR 500	Public
KB6YNO	14063.9	PACTOR I/II	Public
KB6YNO	14108.9	PACTOR I/II/III	Public
KB6YNO	18098.9	PACTOR I/II	Public
KC4TVO	3595.0	PACTOR I/II/III	Public
KC4TVO	7103.7	PACTOR I/II/III	Public
KC4TVO	10139.5	PACTOR I/II/III	Public
KC4TVO	14106.7	PACTOR I/II/III	Public
KC9GQR	10125.5	PACTOR I/II/III	Public
KC9GQR	10132.5	PACTOR I/II/III	Public
KC9GQR	10145.5	PACTOR I/II/III	Public
KC9GQR	14086.5	PACTOR I/II/III	Public
KD6OAT-5	7086.0	WINMOR 500	Public
KD6OAT-5	10140.0	WINMOR 500	Public
KE7XO	3587.0	PACTOR III	Public
KE7XO	7083.0	PACTOR I/II	Public
KE7XO	7103.0	PACTOR III	Public
KE7XO	10142.0	PACTOR III	Public
KE7XO	10147.0	PACTOR III	Public
KJ5YN-5	7097.5	WINMOR 500	Public
KK5AN	3589.0	PACTOR III	Public
KK5AN	7103.4	PACTOR III	Public
KK5AN	10141.2	PACTOR III	Public
KK5AN	14098.7	PACTOR III	Public
KL7EDK	3595.0	PACTOR I/II/III	Public
KL7EDK	7065.9	PACTOR I/II	Public
KL7EDK	7104.4	PACTOR III	Public
KL7EDK	10147.7	PACTOR I/II/III	Public
KL7EDK	14064.0	PACTOR I/II	Public

Call Sign	Frequency (kHz-USB)	Mode	Access
KL7EDK	14098.5	PACTOR III	Public
KQ4ET	3589.0	PACTOR I/II/III	Public
KQ4ET	7067.9	PACTOR I/II	Public
KQ4ET	7101.2	PACTOR I/II/III	Public
KQ4ET	10146.5	PACTOR I/II/III	Public
KQ4ET	14110.0	PACTOR I/II/III	Public
KQ4ET	18106.9	PACTOR I/II/III	Public
KQ4ET	21098.7	PACTOR I/II/III	Public
LA3F	7048.5	PACTOR I/II	Public
LA3F	7052.0	PACTOR I/II/III	Public
LA3F	7054.5	PACTOR I/II/III	Public
LA3F-5	3595.0	WINMOR 500	Public
LA3F-5	597.9	WINMOR 500	Public
LA3F-5	3607.0	WINMOR 1600	Public
LZ1PKS	7043.5	PACTOR I/II/III	Public
LZ1PKS	10139.5	PACTOR I/II	Public
LZ1PKS	10145.9	PACTOR III	Public
LZ1PKS	14111.9	PACTOR I/II/III	Public
NØIA	3587.2	PACTOR I/II	Public
NØIA	3590.0	PACTOR I/II/III	Public
NØIA	7063.9	PACTOR I/II	Public
NØIA	10133.9	PACTOR I/II	Public
NØIA	14074.9	PACTOR I/II	Public
NØIA	14098.7	PACTOR I/II/III	Public
NØIA	14113.5	PACTOR I/II/III	Public
NØIA	18106.2	PACTOR I/II/III	Public
NØIA	21074.9	PACTOR I/II/III	Public
N1ICS	3591.0	PACTOR I/II/III	Public
N1ICS	7076.9	PACTOR II	Public
N1ICS	7101.2	PACTOR III	Public
N1ICS	10141.2	PACTOR II/III	Public
N1ICS	14075.9	PACTOR I/II	Public
N1ICS	14106.0	PACTOR III	Public
N7NMS-5	3589.0	WINMOR 500	Public
N9LOH-5	3579.2	WINMOR 500	Public
N9LOH-5	7096.5	WINMOR 500	Public
N9LOH-5	14106.5	WINMOR 500	Public
NJ7C-5	7089.5	WINMOR 1600	Public
NP2E	7011.4	PACTOR I/II/III	Public
NP2E	10145.0	PACTOR I/II/III	Public
OE3XEC	3608.5	PACTOR I/II/III	Public
OE3XEC	3617.5	PACTOR I/II/III	Public
OE3XEC	10146.5	PACTOR I/II/III	Public
OE4XBU	14064.9	PACTOR I/II/III	Public
OE4XBU	14074.9	PACTOR I/II/III	Public
OE4XBU	14110.0	PACTOR II/III	Public
OE4XBU	21074.9	PACTOR II/III	Public

Call Sign	Frequency (kHz-USB)	Mode	Access
OE4XBU	21098.0	PACTOR II/III	Public
OE4XBU	21117.9	PACTOR II/III	Public
OE5XIR-5	3605.5	WINMOR 1600	Public
PA1JLG-5	14114.5	WINMOR 500	Public
PA3DUV	3583.5	PACTOR II/III	Public
PA3DUV	3593.5	PACTOR II/III	Public
PA3DUV	10136.9	PACTOR II/III	Public
PA3DUV	10141.0	PACTOR II/III	Public
RT9K	14104.7	PACTOR I/II/III	Public
RT9K	14110.5	PACTOR I/II/III	Public
S51SLO	3644.0	PACTOR II/III	Public
S51SLO	7044.0	PACTOR II/III	Public
S51SLO	14144.0	PACTOR II/III	Public
S51SLO	18144.0	PACTOR II/III	Public
UA6DX-5	10145.5	WINMOR 1600	Public
UA6DX-5	14098.5	WINMOR 1600	Public
VA3IED-5	7097.5	WINMOR 1600	Public
VA7DEP-5	7088.5	WINMOR 1600	Public
VE2AFQ	7094.0	PACTOR I/II/III	Public
VE2AFQ	10137.9	PACTOR I/II	Public
VE2AFQ	14068.9	PACTOR I/II/III	Public
VE3AWN-5	7100.0	WINMOR 1600	Public
VK2HL-5	7040.0	WINMOR 1600	Public
VK2JN-5	3631.5	WINMOR 1600	Public
VK2JN-6	14112.0	WINMOR 1600	Public
VK2SYD	3620.2	PACTOR I/II/III	Public
VK2SYD	7046.7	PACTOR I/II/III	Public
VK2SYD	10116.2	PACTOR I/II/III	Public
VK2SYD	14106.7	PACTOR I/II/III	Public
VK2SYD	18107.0	PACTOR I/II/III	Public
VK2SYD	21298.7	PACTOR I/II/III	Public
VK3PG	7068.3	PACTOR I/II/III	Public
VK3PG	7098.5	PACTOR I/II/III	Public
VK3PG	10140.0	PACTOR I/II/III	Public
VK3PG	14127.5	PACTOR I/II/III	Public
VK3PG	18126.5	PACTOR I/II/III	Public
VK6KPS	3624.3	PACTOR I/II/III	Public
VK6KPS	7043.5	PACTOR I/II/III	Public
VK6KPS	10135.4	PACTOR I/II/III	Public
VK6KPS	14097.5	PACTOR I/II/III	Public
VK6KPS	18113.8	PACTOR I/II/III	Public
VK6KPS	21126.5	PACTOR I/II/III	Public
VU2GMN	3643.0	PACTOR I/II/III	Public
VU2GMN	7053.0	PACTOR I/II/III	Public
VU2GMN	10118.0	PACTOR I/II/III	Public
VU2GMN	14124.0	PACTOR I/II/III	Public
VU2GMN	18124.0	PACTOR I/II/III	Public

Call Sign	Frequency (kHz-USB)	Mode	Access
VU2GMN	21183.0	PACTOR I/II/III	Public
VU2GMN	24939.0	PACTOR I/II/III	Public
W5SEG-5	3584.5	WINMOR 500	Public
W5SEG-5	7091.5	WINMOR 500	Public
W6IM	7073.9	PACTOR I/II	Public
W6IM	10141.2	PACTOR III	Public
W6IM	14077.0	PACTOR I/II	Public
W6IM	14098.7	PACTOR III	Public
W7IJ	3591.0	PACTOR I/II/III	Public
W7IJ	7068.9	PACTOR I/II	Public
W7IJ	7103.7	PACTOR III	Public
W7IJ	10139.5	PACTOR I/II/III	Public
W7IJ	14069.4	PACTOR I/II	Public
W7IJ	14110.0	PACTOR III	Public
W7ODN	3593.0	PACTOR I/II/III	Public
W7ODN	7069.5	PACTOR I/II	Public
W7ODN	7103.5	PACTOR II/III	Public
WA7ODN-5	3569.5	WINMOR 500	Public
WBØTAX	3595.0	PACTOR II/III	Public
WBØTAX	7103.7	PACTOR III	Public
WBØTAX	10143.7	PACTOR III	Public
WBØTAX	14096.2	PACTOR III	Public
WBØTAX	18106.2	PACTOR III	Public
WB9FHP-5	3576.5	WINMOR 500	Public
WB9FHP-5	7076.5	WINMOR 500	Public
WD8ARZ-5	24927.5	WINMOR 1600	Public
WG3G	3569.0	PACTOR I/II/III	Public
WG3G	7036.9	PACTOR I/II	Public
WG3G	7107.0	PACTOR III	Public
WG3G	10138.0	PACTOR II/III	Public
WG3G	14112.5	PACTOR II/III	Public
WG3G	18101.9	PACTOR II	Public
WG3G	18106.5	PACTOR III	Public
WL7CVG	3589.0	PACTOR I/II/III	Public
WL7CVG	7075.9	PACTOR I/II	Public
WL7CVG	7101.7	PACTOR III	Public
WL7CVG	10143.7	PACTOR I/II/III	Public
WL7CVG	14065.9	PACTOR I/II	Public
WL7CVG	14096.2	PACTOR III	Public
WX4J	3593.0	PACTOR I/II/III	Public
WX4J	7066.9	PACTOR I/II	Public
WX4J	10143.4	PACTOR II/III	Public
WX4J	14066.9	PACTOR I/II	Public
YBØAJZ	3696.5	PACTOR I/II/III	Public
YBØAJZ	7026.5	PACTOR I/II/III	Public
YBØAJZ	10146.0	PACTOR I/II/III	Public
YBØAJZ	14080.4	PACTOR I/II/III	Public

Call Sign	Frequency (kHz-USB)	Mode	Access
YBØAJZ	18105.5	PACTOR I/II/III	Public
YBØAJZ	21134.5	PACTOR I/II/III	Public
YBØAJZ	24925.0	PACTOR I/II/III	Public
ZL2ABN	3639.0	PACTOR I/II/III	Public
ZL2ABN	7030.5	PACTOR I/II/III	Public
ZL2ABN	10143.5	PACTOR I/II/III	Public
ZL2ABN	14084.3	PACTOR I/II/III	Public
ZL2ABN	18131.5	PACTOR I/II/III	Public

featured Internet-to-HF gateway system known as *Winlink 2000* or "WL2K." Jim Corenman, KE6RK, concurrently developed software called *RMS Express*, which is the end-user portion of *Winlink 2000*. Winlink 2000 is a network of participating stations known as RMS radio gateways, all connected to a central server that functions as the "hub" for connectivity to Internet e-mail and position reporting.

Thank to these advancements, an HF digital operator at sea, for example, can now connect to a Winlink 2000 participating network station using *RMS Express* software and their PACTOR controller and exchange Internet e-mail with non-ham friends and family. He can also exchange messages with other amateurs by using the Winlink 2000 network stations as a traditional global "mailbox" operation.

Most Winlink 2000 participating stations scan a variety of HF digital frequencies on a regular basis, listening on each frequency for about two seconds. By scanning through frequencies on several bands, the Winlink 2000 stations can be accessed on whichever band is available to you at the time.

Using *RMS Express*, Winlink 2000 features include...

The Winlink 2000 network is comprised of RMS radio gateway stations on various HF frequencies throughout the world.

• Text-based e-mail with binary attachments such as DOC, RTF, XLS, JPG, TIF, GIF, BMP, etc.

• Position inquiries accessible from both the Internet graphically via APRS and YotReps, e-mail or radio to track the mobile user.

• Graphic and text-based weather downloads from a list of over 400 weather products, covering the entire globe.

• Pickup and delivery of e-mail regardless of the participating station accessed.

• End-user control of which services and file sizes are transmitted from the participating stations, including the ability for each user to re-direct incoming e-mail messages to an alternate e-mail address.

• The ability to use the Internet via Telnet instead of a radio transmission.

• The ability to use any Web browser to pick up or deliver mail over Winlink 2000.

Hardware and Software Requirements

In addition to your HF SSB transceiver, you'll need a PACTOR controller that is capable of communicating in binary mode with either PACTOR I, II or III. At the time this book was published, only five processors met this requirement:

• SCS PTC-II (in all its variations)
• Kantronics KAM+ or KAM 98
• Timewave/AEA PK-232

RMS Express software is required to access Winlink 2000. You can download *RMS Express* free of charge at **www.winlink.org/ClientSoftware**. It contains a spell-checker and has the look and feel of an e-mail "agent" such as *Outlook Express*.

RMS Express will handle all the communication with your PACTOR controller; you don't need to have a separate piece of software running on your computer. If your transceiver is capable of computer (CAT) control, *RMS Express* will even set your frequency automatically to match the frequency of the Winlink 2000 radio gateway station you are trying to reach.

Accessing a Winlink 2000 Station

Information about getting started with Winlink 2000 may be found at **www. winlink.org/GetStarted** or within the *RMS Express* help files.

In a nutshell, you basically compose your message just as you would in Internet e-mail. When you click the **SEND** button, *RMS Express* will access your PACTOR controller and attempt to connect to the Winlink 2000 station. If the connection fails, it will try again later or give you the option of choosing another station.

When you finally connect, *RMS Express* once again behaves just like an

Internet e-mail client. It will upload your message and then "ask" the Winlink station if you have any mail waiting for you. If so, it will download your mail automatically.

Remember that Winlink 2000 stations usually scan through several frequencies. If you can't seem to connect, the Winlink 2000 station may already be busy with another user, or propagation conditions may not be favorable on the frequency you've chosen. Either try again later, or use the built-in propagation feature to connect on another band.

Sending E-mail To and From the Internet

From the Internet side of Winlink 2000, friends and family can send e-mail to you just as they would send e-mail to anyone else on the Net. In fact, the idea of Winlink 2000 is to make HF e-mail exchanges look essentially the same as regular Internet e-mail from the user's point of view. Internet users simply address their messages to **<your call sign>@winlink.org**. For example, a message addressed to **wb8imy@winlink.org** will be available to me when I check into *any* Winlink 2000 station.

Through the *RMS Express* software the ham user can address messages to non-hams, or to other hams, for that matter, by using the same format used in any other e-mail program.

Although Winlink 2000 supports file attachments, remember that the radio link is very slow (especially compared to the Internet). Sending an attachment of more than 40,000 bytes is usually not a good idea. Text, RTF, DOC, XLS, JPG, BMP, GIF, WMO, GREB and TIF files are permitted as long as they are small enough to comply with the particular user-set limit. Where possible, Winlink 2000 compresses files. For example, a 400,000 byte BMP file may be compressed to under 15,000 bytes. However, this is an exception rather than the rule.

WINMOR

A discussion of the Winlink 2000 network would be incomplete without mentioning *WINMOR*. Unlike the PACTOR controllers we discussed in this chapter, *WINMOR* is indeed sound-card based. This relatively new software offers a solution to those who want to access the Winlink 2000 network but don't own a PACTOR controller. *WINMOR* works perfectly with *RMS Express* as the Winlink e-mail client.

WINMOR will operate with just about any sound-device-based station and SSB transceiver. If you are set up to operate PSK31, you are also able to run *WINMOR* by simply downloading and installing the software. See the *WINMOR* page on the Winlink 2000 website at **www.winlink.org/WINMOR**.

In most digital modes the transmitting station sets the speed and the receiving

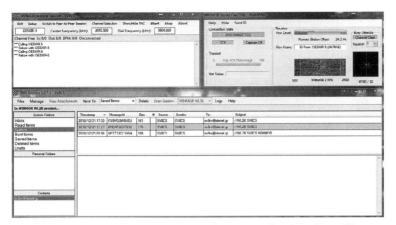

WINMOR is a sound-device-based software application that will allow you to access the Winlink 2000 network on HF without the need for a PACTOR controller.

station must set its speed to match. In *WINMOR* it is the receiving station that determines the speed and lets the transmitting station know what speed it is accepting automatically. The speed you set in setup only comes into effect if someone calls your station in peer-to-peer mode. One of the reasons for this is the RMS stations operating unattended on Amateur Radio frequencies are limited to a 500 Hz bandwidth. MARS stations are not subject to the same restriction so all the Winlink MARS stations will accept a 1600 Hz bandwidth when using *WINMOR*.

Within each bandwidth there are a number of different modes that may be operating. These choices are not under the operator's control and are determined by the signal strength, propagation conditions, multipathing, noise and interference and the bandwidth set by the receiving station. In other words, the program shifts gears to get the best throughput under changing conditions.

Zen and the Casual RTTY Contester

The Short Scoop
Winning isn't everything, grasshopper.
Just playing the game is
its own reward.

Steve Ford, WB8IMY

With a modest station and a hunk of wire in a backyard tree, my odds of placing in the Top Ten of an Amateur Radio contest are comparable to my odds of walking on the Moon – not impossible, but highly unlikely. So, why bother with contesting at all?

The answer is found in the maxim that the journey is often more important than the destination. If you've ever spent an afternoon fishing, you know what it is like to cast your line into the water and watch the bobber as it floats serenely on the surface. You become one with the bobber and the water – a true Zen experience.

Suddenly the bobber plunges into the murky depths and the line goes taut. It's a bite! You grab the pole with your heart pounding and start reeling in...an algae-encrusted twig. Okay, it wasn't an award-winning bass after all, but for a few precious seconds you savored the exciting possibility. You unhook the twig and cast your line again. Maybe next time.

The thrill of contesting is very much the same. A Zen contest master knows that the reward is not in the win, but in the pursuit. The reward is found in pushing the limits of your physical and mental stamina to achieve whatever goal you've set for yourself – even if that goal is defined as making 50 contacts while re-roofing your doghouse. You may never achieve the Olympian summit of the

Top Ten, but you're always a "winner" nonetheless.

I am what is known as a *casual contester*, at the "grasshopper" end of the Zen scale. We are legion in the Amateur Radio contesting world. As much as we'd love to spend an entire weekend doing battle with our radios, life makes other demands. There are lawns to cut, homes in need of repair, families in need of our attention and so on. We have to fit our contesting enjoyment into our otherwise busy schedules and do it in a way that is compatible with everything else going on around us. When life becomes a domestic juggling act, RTTY looks awfully attractive from a contester's point of view.

The Tao of RTTY

With the possible exception of CW, RTTY is one of the oldest digital operating modes in Amateur Radio. In simple terms, you communicate with RTTY by generating a radio signal that shifts between two frequencies at a rapid rate. This is what gives a RTTY signal its distinctive *blee-blee-blee* rhythm. At the receiving end, the shifting signal is decoded into letters, numbers and a limited punctuation set.

In the good old days, RTTY operators used *terminal units* to generate the transmitted signals and decode the received signals. Some hams still use terminal units, but the majority have switched to RTTY software that uses the computer sound card to perform the same task.

Although it isn't as narrow as a CW or PSK31 signal, RTTY still manages to concentrate energy within less than 500 Hz of spectrum, which enhances its ability to be received and decoded when band conditions are marginal. This is why RTTY remains popular among digital DX hunters and contesters.

RTTY isn't the fastest digital mode, but it is reasonably peppy, streaming

Operating the 2003 CQ RTTY Contest using *WriteLog* software. Notice my exchange with LR1F (I've added notes in the RTTY window).

along at about 45 words per minute. Brevity is the soul of wit in a contest – and so is speed.

The fact that modern RTTY is a computer-based mode makes it highly attractive to the casual contester. All your tools are gathered in one place. The computer not only sends and receives RTTY, with the proper software it can maintain your contest log, too. And thanks to *macros* – pre-programmed tasks that your computer executes with the press of a single key, such as responding to a CQ – you can become a versatile multitasker, hunting RTTY contest contacts and doing other things with your computer at the same time. (I once checked my daughter's homework while participating in a RTTY contest.)

What Do You Need to Get Started?

The shopping list for a casual RTTY contest station is remarkably short:
- An HF SSB transceiver
- A computer with a sound card
- A sound-card-to-transceiver interface
- Software

My guess is that you already have most of the above or you wouldn't have read this far. Of the remaining items, the sound card interface is the hardware that allows you to shuttle audio between your radio and your computer. It also handles transmit/receive switching. You can buy an interface from *QST* advertisers such as MFJ Enterprises (**www.mfjenterprises.com**), West Mountain Radio (**www.westmountainradio.com**) or TigerTronics (**www.tigertronics.com**). You can also make your own.

The software can be just about anything that will use your sound card to do the tricks necessary to transmit and receive a RTTY signal. A partial list is shown in the "RTTY Software" sidebar. Some software titles are free.

Strictly speaking, you don't need contest-logging software, but it sure makes life easier. Not only does the software log all your contacts, it automatically checks for duplicate contacts, better known as *dupes*. Depending on the rules of the contest, you may only be allowed to contact the same station once, or once per band. Duplicate contacts are a waste of time and should be avoided.

There are contest-logging software packages that include the ability to send and receive RTTY. Others allow you to incorporate standalone RTTY programs. Browse the web and the advertising pages of *QST* and you'll find that you have many choices.

I could devote several pages describing how to set up a RTTY station, but there is an excellent resource that is as close as your computer keyboard. RTTY contest master Don Hill, AA5AU, has a superb web page that will give you all the detail you need. Get on the web and go to: **www.aa5au.com/rtty.html**. Also,

RTTY Software

This is only a partial list. See the advertising pages of *QST* or search the web for more.

Windows
RCKRTTY: www.rckrtty.de/
WriteLog: www.writelog.com
MMTTY: hamsoft.ca/pages/mmtty.php
TruTTY: www.dxsoft.com/
MixW: mixw.net

Mac
Multimode: www.blackcatsystems.com/
  software/multimode.html
cocoaModem:
  www.w7ay.net/site/Applications/cocoaModem/index.html

Linux
LinPSK (RTTY and PSK31): linpsk.sourceforge.net/
Fldigi: www.w1hkj.com/Fldigi.html

the manuals that accompany commercial sound card interfaces offer guidance on how to set up your station.

It's important to mention that once you've established your RTTY station, you can use the same hardware to operate PSK31, MT-63, Hellschreiber, slow-scan television and many other modes. All you have to do is load a different program. That's a powerful incentive to get started!

Big Guns vs Little Pistols

With Big Gun contest stations, *big* is the operative word – big antennas and big power. We're talking towers and kilowatts. Some contest operations consist of several operators working several transceivers at the same time. This is the *multi-multi* juggernaut. If you ever have an opportunity to join a multi-multi team, I highly recommend it. The team atmosphere is intoxicating. It's more fun than a ham should be allowed to have.

What about the rest of us? We're the Little Pistols (or Little Pop Gun, in my case). Believe it or not, we are highly valuable to the Big Guns despite our reduced profiles. Can you guess why? The answer is that every station, no matter how small, represents a point in a contest. And if you are lucky enough to live in a place that has few RTTY operators in residence, you are even more desirable as a rare *multiplier*.

The Value of Multipliers

Every contest has multipliers. These are US states, DXCC entities, ARRL sections, grid squares and so on, depending on the rules of the contest. A multiplier is valuable because it multiplies your total score.

Let's say that DXCC entities are multipliers for our hypothetical contest. You've amassed a total of 200 points so far, and in doing so you made contacts with 50 different DXCC entities.

$200 \times 50 = 10,000$ points

Those 50 multipliers made a huge difference in your score! Imagine what the score would have been if you had only worked 10 multipliers?

If given a choice between chasing a station that won't give me a new multiplier and pursuing one that *will*, I'll spend much more time trying to bag the new multiplier. So will the Big Guns.

Hunting and Pouncing vs Running

The common sense rule of thumb is that a Little Pistol station should only hunt and pounce. That means that you patrol the bands, watching for "CQ TEST" on your monitor and pouncing on any signals you find. The Big Guns, on the other hand, often set up shop on clear frequencies and start blasting CQs. If conditions are favorable, they'll be hauling in contacts like a trawler. This is known as *running*.

In many cases, Little Pistols are probably wasting valuable time by attempting to run. There are situations, however, where running *does* make sense. If you've pounced on every signal you can find on a particular band, try sending a number of CQs yourself to catch some of the other pouncers. If you send five or 10 CQs in a row and no one responds, don't bother to continue. Move to another band and resume pouncing.

You can also run successfully if you are a rare multiplier. No matter how weak your signal may be, all the other stations will come to *you!*

As you make each contact, keep your exchanges short. The goal is to communicate the required information clearly. Anything else is extraneous. Let's say that you're involved in a contest where the object is to exchange a signal report and state…

[RIGHT]   K1RO   599   CT   CT   DE   WB8IMY   K

WB8IMY gives K1RO a 599 signal report (they are all "599" – one of the peculiarities of contesting) and then repeats his state abbreviation twice (CT – Connecticut) before quickly ending the transmission.

[W R O N G]   K1RO DE WB8IMY...Thanks for the contact. You are 599 here in the state of Connecticut. Weather here is wonderful. My wife is divorcing me and my cat is coughing up a hairball in my lap. K1RO DE WB8IMY K

This is more information than K1RO is likely to want – even outside of a contest! In contesting, time is critical. Say what you have to say clearly and *quickly*. In addition, the longer you transmit, the more opportunities there will be for errors caused by fading, noise or interference.

The rewards of casual contesting, like fishing, are in the eye of the beholder, or the Zen master. A few dozen contacts to test the waters may suffice. Or you may choose to bang out as many contacts as your time allows. In the end you'll take home a collection of soggy branches (minus some hooks and bobbers), or you might land the mother of all muskies. Either way, you'll leave fulfilled, at one with the contest universe.

Ghost QSOs –
Olivia Returns from the Noise

The Short Scoop
Olivia – the magic mode.

Gary L. Robinson, WB8ROL

It was a chilly foreboding fall evening when I fired up the Yaesu FT-100D, started my DM780 software (part of the *Ham Radio Deluxe* suite) and parked my dial frequency on 1.808 MHz. The waterfall was as empty and devoid of signals as a sunspot-starved ionosphere. After listening for almost half an hour without hearing any signals, I decided to set the program up to call CQ automatically once a minute.

I set my mode as Olivia 500/16 (500 Hz wide using 16 tone format) – a digital mode I had recently "discovered" and was intrigued with – and double-checked the frequency to make sure it was not in use. I engaged the auto-CQ and reached for my low-G Irish whistle – another favorite endeavor of mine – and commenced playing a short lively jig. I routinely play tunes when sending auto-CQs and keep the receiver audio at a low level. With one ear on the receiver, one eye on the waterfall and both hands on the whistle I multitasked merrily.

I had not heard all that much Olivia digital mode in use on 160 meters (or on the other bands) so I often found that calling CQ in that mode was more effective than just listening for activity. As I was slogging my way through O'Carolan's Concerto for the second time, whistling like a piper on fire, the auto-CQ already had finished five calls. So far I had not heard the cacophony of tones signifying an Olivia signal or noticed anything at all in the waterfall window. It was at this moment that I entered the world of the paranormal.

I chanced to glance at the text box that displayed received (decoded) text and I saw characters appearing. My call letters appeared as if a ghostly presence was typing them and answering my CQ. I screeched to a halt on my whistling, immediately disabled the auto-CQ and turned up the receiver volume. With the volume all the way up, all I could discern was the usual S8 hiss and static typical of nights on 160 meters. The waterfall still did not show any signal at all – but the characters kept appearing! What was going on here? Was it time to shut the rig off and consider bed rest and rehabilitation? Was it a bad piece of meat or an uncooked portion of potato I had for dinner that was causing hallucinations? Or should I attempt to answer the ghostly call?

There's No 0 for Signal Strength

Well, I am a ham first and foremost so I cautiously clicked on the transmit button and began to tap out my reply – only hesitating when I had to consider just what I should give as a signal report. I finally banged out a 519 RST (Readability, Signal Strength, Tone) report (only because there is no zero for signal strength) and sent it back to my ghostly friend to see what would happen next.

Well, we went back and forth for several transmissions as we exchanged all the usual information and then we ended the QSO passing a friendly 73 to each other. My ghostly friend appeared to be a real ham in Texas and checked out okay on the web site **www.QRZ.com**. My sanity was preserved and apparently intact (my wife doesn't totally agree on this point), though I was still slightly confused and filled with wonderment by what had just transpired!

Since that day I have had several more of what I refer to as "Ghost" QSOs using the Olivia mode and have done a little online research to reassure myself that this is not all that uncommon. For the previous three and a half years I had mostly operated on PSK31 (a digital mode using phase shift keying at 31.25 bits/sec) and a little MFSK16 (Multi Frequency Shift Keying; a digital mode using 16 tones) and yet I had never experienced this type of ghostly phenomenon with either of these excellent digital modes. Apparently Olivia is one of the few modes that can actually decode signals that are at or below the noise level. That is why I *never* use the squelch with this mode because it could cut off signals below the noise floor that would otherwise be decoded.

The Internet resource Wikipedia.org says, "Olivia MFSK is an Amateur Radio teletype protocol developed by Pawel Jalocha, SP9VRC, in late 2003 to work in difficult (low signal-to-noise ratio plus multipath propagation) conditions on shortwave bands. A signal can still be properly copied when it is buried 10 dB below the noise floor (ie when the amplitude of the noise is slightly over three times that of the signal)." The Wikipedia entry includes several paragraphs describing the technical details of the mode.[1]

Now, I am neither a technical whiz nor an expert in any field but I found myself smitten. I don't have a giant tower with humongous beams, run high power or even live at a great ham location, but Olivia in particular has made it possible for me to chase DX and have quality ragchews like no other mode I have ever used. It has really ignited my activity level and fun factor to heights that rival my early Novice days and later DX chasing when I did have a big tower and massive antennas.

Let Me Introduce You

Olivia mode can be set to various formats that are labeled using the particular format's bandwidth and number of tones. Bandwidths of 125, 250, 500, 1000 and 2000 Hz are typical. The number of tones can be set anywhere from 4 to 256 depending upon the propagation conditions. Different combinations of tones and bandwidths provide for slower or faster transmission rates. Commonly used formats are 125/4 (125 Hz bandwidth using 4 tones), 250/8, 500/16, 500/8 and 1000/32. The 500/8 format seems most popular at this moment, though I have had several QSOs on the 125/4 and 250/8. The 1000/32 format seems to be popular on 20 meters.

Each of the Olivia formats has advantages and disadvantages. Obviously, the bandwidth differences make the more narrow formats attractive because they will fit in available open spectrum space more easily. They also will likely get through slightly better since all the power of the transmitted signal is concentrated in a smaller bandwidth – much the way CW gets through better than wide phone signals.

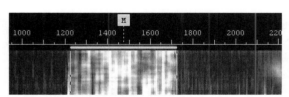

Close-up of an Olivia 500/16 *average* format transmission shown on the DM780 waterfall display.

The speed of Olivia is an issue also. Olivia is generally not as fast as PSK31 or MFSK16. Olivia 500/16 sends text at approximately 20 wpm. The 500/8 format speeds that up to nearly 30 wpm. Fewer tones results in more speed while less bandwidth results in slower speed. Olivia 1000/8 and 2000/8 are often used by Military Auxiliary Radio System (MARS) traffic nets because these formats are fairly fast, accurate and get through when the MT63 mode (a digital mode using 64 tones phase-shift-keyed in a 1 kHz bandwidth) fails. Most of this information and more pertaining to Olivia are available at the HFLink and DXZone web sites.[2,3]

Many hams find Olivia slower than they like and prefer to use other modes,

Olivia QSO Formats

Format Bandwidth/Tones	Audio Center Marker (Hz)	Baud	Decode S/N Ratio (dB)	Speed WPM
500/16*	750	31.25	−13	19.5
1000/32*	1000	31.25	−12	24.4
500/8	750	62.5	−11	29.3
1000/16	1000	62.5	−10	39.1
500/4	750	125	−10	39.1
250/8	625	31.25	−14	14.6

*The most common Olivia formats in use at this time.

while many others find the accuracy and ability to get through an acceptable trade-off. Also, many of us, including myself, are not fast typists and actually find "slower" to be a positive attribute and allows for more comfortable overall operation.

Another advantage to the mode is that it's not quite as critical for it to be tuned exactly on frequency as it is with PSK, MFSK and many other modes. If you click on the waterfall with your mouse and the indicator doesn't get exactly on the signal it may still decode properly. Most implementations of Olivia are set to search for signals to either side of your center frequency by a fixed percentage of your signal's width.

Pioneers Wanted

The activity level of Olivia can be described as somewhat sparse, especially compared to PSK and CW. That is partly explained by the fact that only a few programs support Olivia at this time and that Olivia has not been around quite as long as other established modes. *MixW* (newer versions), *MultiPSK*, *Ham Radio Deluxe's DM780* (latest Beta versions)[4] and *OliviaMFSK* support Olivia operation and are available for *Windows*-based computers, while the *Fldigi* (v3) program is available for *Linux* and *Windows XP/Vista*.

There are a few other things worth mentioning about Olivia. Since it does incorporate some error correcting, it is not totally a "real time" mode. It is essentially a real time mode in the sense that you do not "connect" to a station and it is not duplex- or bulletin-board (BBS) type operation.

Like PSK, you go back and forth in a typical ham QSO fashion. The difference is when you click on the waterfall of a PSK signal it starts decoding very quickly. When you do the same on an Olivia signal it may take 3 to 6 seconds before it starts decoding. The opposite happens at the end of a transmission. When the station you are listening to stops transmitting – your Olivia software program will continue to decode for three to six seconds. Typically that will result in a six to 12 second gap when passing transmissions back and forth.

This varies depending on the format and the software implementation in the program you are using. I see less of a gap using *Fldigi* than with *DM780* for instance. What this means is that when you click on a signal if you don't see anything right away – be patient! It takes a few seconds. When you send it back to the other station – again, be patient. It may be a few seconds before you start to see his printout or even hear him.

The QSO is on the Calling Frequency

Another aspect of Olivia operation that is slightly different from most other digital modes is a loose voluntary channelization of the frequencies used by many operators. Since it is possible to copy signals that you cannot hear or that you might not see on the waterfall, it makes sense to use specific designated frequencies for calling and meeting other stations. Otherwise, if you just tuned all over the place looking or listening for an Olivia signal you could miss a lot of stations.

A few of the popular frequencies are 14.107.50 MHz (20 meter calling frequency 1000/32 format generally), 7072.50 MHz
(40 meter calling frequency 500/250/125 Hz bandwidth formats generally) and 10.133.65 MHz (30 meter calling frequency 500/250/125 Hz bandwidth formats generally). These frequencies are *dial* settings – meaning the frequency that your transceiver would display on *upper* sideband. For the 500/250/125 Hz bandwidth formats the waterfall position (center marker) should be set for 750 Hz. On the 1000 Hz bandwidth format the waterfall position should be set at 1000 Hz, and with the 2000 Hz bandwidth format it would be set at 1500 Hz.

These "channels" and settings are not cast in stone and are certainly not mandatory or used by all Olivia stations but they are useful especially to help facilitate weak signal QSOs. For a much more complete set of voluntary frequencies see the charts at the HFLink web site.

Better Than CW?

After a full year of operating Olivia I have had 650 QSOs with 45 states and 21 countries and have found it to be more reliable and more fun than any other mode that I have ever used on ham radio – including CW. It gets through noise (QRM), static (QRN) and fading (QSB) better than most and is excellent for DXing, ragchewing and even VHF weak signal QSOs. The faster Olivia formats are very useful for handling message traffic. The only place where it fails to shine as brightly is during contests. With many of the Olivia formats being slower and with the "not quite 100%" real time quality (four to six second delay) it is probably not well suited for contesting.

So in the end, it turns out that a really great communications mode and not ghosts were responsible for my extremely weak and spooky QSOs. I will put away my ghostbusting tools, discard the books about Area 51 and fire up the rig to see if I can scare up some more Olivia activity! Give it a try and discover the magic and mystique of Olivia!

~ Gary L. Robinson, WB8ROL, was first licensed in 1963 as WN8GIG at the age of 13 and obtained his Extra class license in the mid 1980s. He is semi-retired and living in a small town in rural Ohio with his spouse Nancy. During his active work years he wore many hats — including Corrections Officer for 14 years, computer technician and C, C++ and C# programmer. Currently he focuses on his five loves — ham radio (digital, especially Olivia, DominoEX FEC and THOR modes), computers and programming (primarily but not exclusively on Linux), playing Irish whistles, five cats and his wife. An ARRL member, Gary can be contacted at **wb8rol@arrl.net** *or look for him lurking in the digital subbands of 160 meters through 70 cm.*

Notes
[1]http://en.wikipedia.org/wiki/Olivia_MFSK
[2]http://hflink.com/olivia
[3]www.dxzone.com/cgi-bin/dir/jump2.cgi?ID=12012
[4]http://hrd.ham-radio.ch

Reverse Beacons

Steve Ford, WB8IMY

Here in the US, the Federal Communications Commission allows amateurs to establish unattended beacon stations within various frequency ranges beginning at 28.200 MHz. If you take a quick spin through the beacon sub-band on 10 meters, for instance, you'll likely hear the Morse code IDs of several beacons, assuming the band is open.

Reverse beacons turn the traditional beacon concept upside down.

In the land of reverse beacons, there are no unattended transmitting stations, although there may be plenty of unattended *listening* stations. These listening stations monitor the bands for CW and digital signals. The signals may be from stations engaged in conversations, or from stations calling CQ. Whenever a listening station decodes a call sign, it notes the frequency, time and occasionally the signal strength and then forwards the information automatically to a website for everyone to see.

The Reverse Beacon Network (RBN) maintained by Pete Smith, N4ZR, and Felipe Ceglia, PY1NB, depends on volunteer listening stations that use *CW Skimmer* software developed by Alex Shovkoplyas, VE3NEA, to comb the HF bands for CW activity. You'll find the RBN at **www.reversebeacon.net**.

Another network, and one of my personal favorites, is PSKReporter maintained by Phil Gladstone, N1DQ, at **http://pskreporter.info**. Don't let the name of the site fool you. This reverse beacon network is about much more than PSK. Phil's site aggregates reception reports of many different kinds of modes from CW to JT65 and displays the

The Reverse Beacon Network at www.reversebeacon.net.

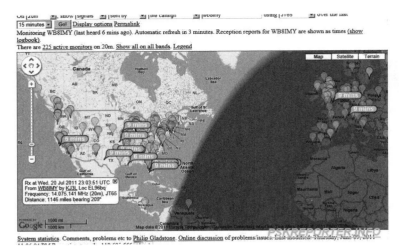

System statistics. Comments, problems etc to Philip Gladstone. Online discussion of problems/issues. Last modified: Thursday, June 09, 2011

After making a contact on 20 meters using JT65 with just 1 W to a mobile antenna, I quickly jumped to the PSKReporter site and saw that a surprising number of other stations had heard me as well.

The WSPRnet propagation map.

results on a near-real-time map. There are dozens of listening stations contributing reports to PSKReporter at any given moment.

I enjoy operating JT65 on the HF bands and use *JT65-HF* software by Joe Large, W6CQZ. It is utterly fascinating to try different antennas or power levels and see the results on PSKReporter. I can make a contact or call CQ and within 60 seconds I'll see reports on the network. A glance at the map tells me which stations heard me and where they are located. Joe maintains his own JT65 reverse beacon network as well at **http://jt65.w6cqz.org/receptions.php**.

WSPRnet is yet another take on the same idea, but this network is built around stations using K1JT's MEPT-JT mode at very low power levels. The reporting website at **http://wsprnet.org/drupal/** also offers an informative map display.

Remember that you don't have to be sitting at your radio to contribute reports to any of these networks. You aren't transmitting; you are simply listening. All you need is an Internet connection and the appropriate software depending on which network you wish to join. I routinely activate the reporting function in *JT65-HF*, park my rig on a JT65 frequency (such as 14.076 MHz) and walk away for hours. While I'm out running errands or puttering around the yard, I'm also rendering a service to my fellow hams. Such a deal!

Digital VOX Sound Card Interface

The Short Scoop
Here's an interface that will get you on the air
with PSK31 and other digital modes regardless of
the type of computer you are using –
no serial or USB cable required!

Howard "Skip" Teller, KH6TY

Most sound card interfaces are powered either by a voltage from the computer serial port, by a voltage taken from the computer accessory port or microphone port, or by a "wall wart" ac adapter. If the computer has no serial port, and most computers these days have USB ports instead of serial ports, it is also necessary to use a USB serial adapter to generate a *virtual* serial port that the communications software can use for push-to-talk operation.

It would be more convenient if no dc voltage were needed to power the interface, and also if no serial port or USB serial adapter were needed, so I wanted to find a way to eliminate both the need for a serial port or USB serial adapter and a dc voltage. I then realized that computer sound cards have evolved from having both a high level speaker output jack and a line-level audio output jack, to usually having just an earphone/headphone jack and a microphone jack. Measuring the maximum audio output level of this jack on several computers, and also on an external sound card, such as "USB sound adapters" commonly sold to provide microphone and earphone jacks via a USB connector, I found that it was generally around 2.5

V peak-to-peak – not enough to power a switching transistor in an interface.

By connecting the earphone/headphone output to the center tap of a 600:600 Ω isolation transformer, however, that voltage is doubled across the full secondary winding of the isolation transformer to 5 V peak-to-peak – enough to rectify and power a transistor switch for push-to-talk switching. Since all this occurs at the secondary winding of the transformer, the transformer still isolates the computer earphone output and ground from the transceiver itself, thereby preventing any hum or ground loops from disturbing the transmit or receive audio.

By using another isolation transformer for the receive audio, the computer is totally dc-isolated from the transceiver, both on the audio input lines for transmit, the receive audio output line for receive, and the push-to-talk switching line for transmit/receive switching. The schematic diagram in Figure 1 shows how this isolation is provided by the transformers.

How It Works

See **Figure 1**. Sound-card-based digital communications software, such as *DigiPan*, generates a WAV audio signal when placed in the transmit mode. This WAV audio contains the modulation that you use to communicate with on digital modes, such as PSK31 and others, and is used to modulate the transceiver in the same way that you modulate the transceiver audio with the microphone when operating SSB. Since this WAV audio comes out of the earphone or speaker jack of the computer, that jack is connected to isolation transformer T1, but only between one side of the primary winding and the center tap of the primary winding. When this audio is coupled across the transformer to the full secondary winding, it appears at twice the value that is present between the primary center tap and either end of the primary winding, because the turns ratio of the transformer in that case is no longer 1:1, but 1:2.

The WAV audio is then used to modulate the transceiver through the data, accessory or microphone jacks for digital operating. The audio level is adjusted by means of potentiometer VR1.

In order to switch the transceiver from receive to transmit, this same double-value ac voltage is applied to capacitor C1 and resistor R1, which isolate the transformer audio from the switching action of the following voltage doubler circuit, D1 and D2. These diodes form a classic dc voltage doubler rectifier circuit, which conducts on both positive and negative cycles of the WAV audio. C2 then charges up to the peak value of the ac voltage and holds that charge long enough to drive the base of switching transistor Q1, through current limiting resistor R2, causing the collector of Q1 to saturate and pull the push-to-talk pin

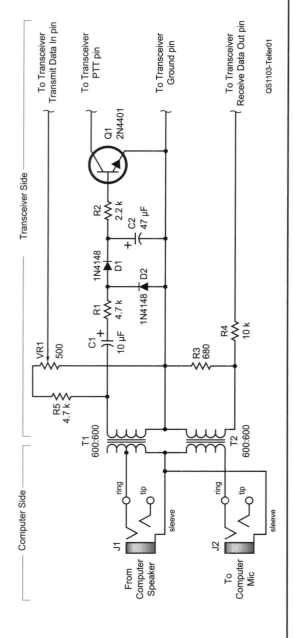

Figure 1 – Schematic diagram of the Digital VOX Sound Card Interface. Vendor part numbers are shown in parenthesis. Components can be obtained from Mouser Electronics, 800-346-6873; www.mouser.com. Parts list on facing page.

Parts List

C1 – 10 µF, 50 V electrolytic capacitor
 (Mouser 647-UVR1H100MDD1TA).
C2 – 47 µF, 50 V electrolytic capacitor
 (Mouser 647-UVR1H470MED1TD).
D1, D2 – 1N4148 diode
 (Mouser 512-1N4148).
J1, J2 – 1/8 inch stereo jack
 (Mouser 161-3507-E).
Q1 – 2N4401 or other switching transistor
 (Mouser 512-2N4401BU).

R1, R5 – 4.7 kΩ, 1/4 W resistor
 (Mouser 291-4.7K-RC).
R2 – 2.2 kΩ, 1/4 W resistor (Mouser 291-2.2K).
R3 – 680 Ω resistor (Mouser 291-680-RC).
R4 – 10 kΩ, 1/4 W resistor
 (Mouser 291-10K-RC).
T1, T2 – 600CT:600CT isolation transformer
 (Mouser 42XL016-RC).
VR1 – 500 Ω potentiometer
 (Mouser 652-3362-1-501LF).

of the transceiver to ground. Q1 gets its operating collector voltage from the push-to-talk circuit in the transceiver, which must be designed, as almost all transceivers are these days, for an "open collector" switch for transmit/receive switching.

Transformer T2 is used to isolate the receive audio to the computer from the transceiver, and the primary is connected to the computer microphone input. The receive audio output voltage is fed from the transceiver through an L-pad consisting of R4 and R3, attenuating the high audio output of the transceiver data jack, or earphone jack, to a suitable level for the microphone input of the computer. It is this input to the microphone of the computer that creates the "waterfall" display of the typical digital communications program that also decodes the WAV audio being transmitted by the other station into characters on the screen. Resistors R4 and R3 can be exchanged in position if the computer audio input requires a higher audio level from the transceiver.

Adjusting the Transmit Level

Before connecting the audio cables, you should be able to hear the WAV audio in the computer speakers when the software is in the transmit mode. Plug in both cables between the computer and transceiver. With the software waterfall cursor around 1500 Hz and the software in transmit, raise the audio output level until the transceiver switches into transmit. Once again, if the software you are using doesn't offer the means to make this adjustment you'll need to access the audio levels within *Windows* Control Panel.

When the transceiver switches into transmit, increase the audio output level about one more notch to ensure there is enough audio for push-to-talk to consistently activate. Adjust the transceiver RF power control for maximum, or if using the microphone jack on the transceiver, to the normal position for SSB phone operation. Now adjust VR1 until the RF output of the transceiver is about 30% of rated maximum power. It should not be necessary to make this adjustment again. Just raise the audio output level control to obtain a little more output power if desired, but at 30% power, you should automatically have a clean PSK31 signal.

After setting the RF power level, place the wired circuit board into the enclosure and snap the two halves together. This completes the assembly and adjustment of the interface. If you find it necessary to adjust the interface very often, it might be convenient to drill a small hole in line with VR1 so a screwdriver can be inserted into the enclosure to adjust VR1. In most cases, however, it should be possible to just set VR1 and leave it alone, doing all the fine power setting adjustments with the software audio level controls. Near the extreme edges of the IF passband, the audio output of the transceiver may decrease because of the shape of the IF filter, so it may be necessary to increase the audio output level to maintain push-to-talk action. Remember to recheck the power output when retuning to the center of the passband.

The digital VOX interface works well with all digital modes such as PSK31 and AFSK RTTY, but does not work correctly with sound card CW or Hellschreiber. If you intend to operate those modes, the Classic Sound Card Interface described on page 37 of the July 2010 *QST* is a better choice because its serial port switching keeps the transceiver in transmit until the software returns it to receive. It is not dependent upon the audio tones being present to keep it in transmit.

~ Howard "Skip" Teller, KH6TY, is an ARRL member and was first licensed in 1954. He received his commercial First Class Radiotelephone license in 1959 and worked his way through college as chief engineer of several radio stations. He holds a BS degree in electrical engineering from the University of South Carolina and is retired from running a factory in Taiwan, where he manufactured the weather-alert radio he originated in 1974 and is still sold by RadioShack and many other companies now. Skip enjoys developing digital software, such as DigiPan *and* NBEMS, *and designing VHF/UHF antennas. He is currently studying the potential of working 432 MHz DX using the Contestia digital mode. You can contact Skip at 335 Plantation View Ln, Mt Pleasant, SC 29464;* **skipteller@gmail. com**.

Notes

Notes

Notes

Notes

Notes

Notes

Index

FEEDBACK

Please use this form to give us your comments on this book and what you'd like to see in future editions, or e-mail us at **pubsfdbk@arrl.org** (publications feedback). If you use e-mail, please include your name, call, e-mail address and the book title, edition and printing in the body of your message. Also indicate whether or not you are an ARRL member.

Where did you purchase this book?
  ☐ From ARRL directly   ☐ From an ARRL dealer

Is there a dealer who carries ARRL publications within:
  ☐ 5 miles   ☐ 15 miles   ☐ 30 miles of your location?   ☐ Not sure.

License class:

☐ Novice

☐ Technician

☐ Technician with code

☐ General

☐ Advanced

☐ Amateur Extra

Name _____

Daytime Phone   (   ) _____

Address _____

City, State/Province, ZIP/Postal Code _____

If licensed, how long? _____

Other hobbies _____

_____

Occupation _____

ARRL member?   ☐ Yes ☐ No
Call Sign _____

Age _____

E-mail_____

For ARRL use only	HFDIG
Edition	1 2 3 4 5 6 7 8 9 10 11 12
Printing	1 2 3 4 5 6 7 8 9 10 11 12

From _____

_____

_____

EDITOR, GET ON THE AIR WITH HF DIGITAL
ARRL—THE NATIONAL ASSOCIATION FOR AMATEUR RADIO
225 MAIN STREET
NEWINGTON CT 06111-1494

— — — — — — — — — — — — — please fold and tape — — — — — — — — — — — — — — —